Plate Photography
Unless otherwise stated, all images are courtesy and copyright © Angela Moore

Peter Guenzel 216–17; Industrial Facility 37 (portrait), 48–50; Gerhardt Kellermann 218–19; Wästberg 84–5

Other photography
Unless otherwise stated, all images are courtesy and copyright © Industrial Facility

Richard Davies 258 (b); Elle Deco 149 (tl), 283 (ml); Domus 141 (bl), 142 (tl, tr); Ben Evans 282 (b), 283 (mr); Maja Flink 285 (tl); Fabian Frinzel 139 (m, bl); Graphic Thought Facility 240 (b), 266 (tl); Peter Guenzel 136 (b); Hello Design 282 (tl), 284 (tr); Herman Miller 119 (t), 151 (br), 152 (b), 153 (t), 220, 274 (tl); Sasaki Hidetoyo 284 (mr); Hitachi 268 (br); Philipp Karlberg 264 (tl); Gerhardt Kellermann 128 (b), 273 (b), 274 (m, bl, br), 277 (bl), 282 (mr), 285 (tl); Leach pottery/Tate Britain 253; Christoph Mack 283 (tr), 284 (ml); Nicole Marnati 273 (m); Angela Moore 140 (br), 154 (b); Muji 257 (ml, mr), 258 (tl); Aleksi Niemela/ Margaret Howell 125 (t); Jara Varela 147 (b); Wästberg 134, 147 (tr), 168–9; Yamaha 271 (t); Miro Zagnoli 150; Max Zambelli 259 (t); Maria Ziegelböck 282 (tr)

Acknowledgements
Industrial Facility would like to thank the many assistants who have worked with us over the years: Kevin Anthoine-Milhomme (France), Shota Aoyagi (Japan), Christian Beck (Germany), Katharina Bellinger (Germany), Gregory Brunisholz (Switzerland), Julia Casanova (Spain), Il Gu Cha (South Korea), Trevor Cheney (USA), Juhee Cho (South Korea), Lihsuan Chou (Taiwan), Moon Chul (Korea), Jessie Colin (USA), Sean Collins (Canada), Luca Corvatta (Italy), Renaud DeFrancesco (Switzerland), Sarha Duquesne (France), Andrew Fallshaw (Australia), Eva Feldkamp (Germany), Joseph Guerra (USA), Keigo Harada (UK), Jon Harrison (UK), Joel Hoff (Germany), Christine Huhn (Germany), Adam Huxley-Khng (UK), Emilia Izquierdo (Portugal), Dustin Jessen (Germany), Florent Julien (France), Sejoon Kim (Korea), Noam Kollman (Israel), Ronan Le Vacon (France), Ippei Matsumoto (Japan), Nicolas Le Moigne (Switzerland), Gwendolyn Nichols (USA), Andrea Pavan (Italy), Klara Peterson (Sweden), Roman Pin (France), Liisa Poskiparta (Finland), Frederic Julian Rätsch (Germany), Jonathan Renous (France), Julian Ribler (Germany), Manuel Roeck (Germany), Keenan Rush (Switzerland), Petra Schmidt (Germany), Gen Suzuki (Japan), Kohei Tajima (Japan), Phan Thao (UK), Philipp von Lintel (Germany), Romain Voulet (France), Adrian Weidmann (Switzerland), Edvin Wester (Sweden), Kukhyun Yoon (South Korea)

Sam Hecht and Kim Colin would also like to extend their gratitude to designers Paul Neale and Carole Courtillé of Graphic Thought Facility for their uncompromising spirit in the making of this book, and to photographer Angela Moore for her physical and mental prowess. We are grateful to all of our clients and especially to Herman Miller, Elephant & Coral, Dinesen, Mattiazzi and Emeco for their support of this book; to the many contributors who have given up their time; to our families; and to Phaidon Press editors Emilia Terragni and Robyn Taylor.

Phaidon Press Limited
Regent's Wharf
All Saints Street
London N1 9PA

Phaidon Press Inc.
65 Bleecker Street
New York, NY 10012

phaidon.com

First Published in 2018

ISBN 978 0 7148 7579 8

A CIP catalogue record for this book is available from the British Library and the Library of Congress.

Commissioning Editor:
Emilia Terragni
Project Editor:
Robyn Taylor
Production Controller:
Alenka Oblak
Design:
Graphic Thought Facility

Art Direction of Plate Photography:
Graphic Thought Facility with
Sam Hecht and Kim Colin

Printed in China

Industrial Facility Sam Hecht and Kim Colin

Industrial Facility Sam Hecht and Kim Colin

Foreword Alain de Botton

Years ago, I came to know Sam Hecht and Kim Colin of Industrial Facility through their Branca chair. Their brilliant minds and talent, which in other eras might have been employed to bring utility to fireplaces or comfort to glassware, are now spent making objects of everyday life with uncommon beauty and grace.

Why *is* design important? Partly because bad design is a symbol of a lack of care and empathy. The badly designed object is one that's forgotten what it's like to be us, that's refusing to see it from our point of view. Design ultimately matters because it offers an invitation to live and *be* a certain way. When we call a chair or a lamp beautiful, really what we're saying is that we like the way of life that it suggests to us: we would like to be the way that it already is. We remain hugely vulnerable to the voices of the objects in our environment. Our inner states are heavily open to influence and we may be as harmed by bad design as we are by moral evil.

Industrial Facility (the name itself is a wonder) constantly pare back, underplay, and yet seduce, enliven and over-deliver. Everything they have designed has reduced matters to essentials and yet managed to find a way to delight as well. They make it look so easy, as great artists often do (the Roman poet, Horace, knew as much: 'the art lies in concealing the art').

The ability to create a welcoming experience in design for another person is a skill that derives from an unusual source: remembering what it is like to be distressed around an object. In their work, Industrial Facility are in touch with what it feels like to be lost – to feel abandoned, frustrated, bored. Everything they make is ready for our use and needs no explanation: it understands us better than we understand ourselves.

Their work will last. We live in a Romantic age, which tends to emphasize what is new and obviously surprising. Industrial Facility care about what is permanent. Along the way, they are also sketching a better future for business, one in which the objects of everyday use are imbued with true quality and workmanship. It's a vision of a dignified, ordinary life.

We are taught to be suspicious of materialism. When people want to identify the root cause of corruption in our age, they generally only need to point the finger at our attachment to material things. But it's never actually materialism – the pure fact of buying things and getting excited by possessions – that's really the problem. Material objects play a positive psychological (or spiritual) role in our lives when higher, more positive, ideals are 'materialized' in them; when buying and using objects on a daily basis gives us a chance to get closer to our better selves. When contained in physical things, valuable psychological qualities, which are otherwise often only intermittent in our thoughts and conduct, can become more stable and resilient.

In their work, Colin and Hecht are like the ideal parents: they make the world friendlier and more hopeful for us, while hiding the sacrifices that they have had to make in order to do so.

Closeness to the World Peter Kapos

Sam Hecht and Kim Colin came together to form Industrial Facility in London in 2002, but only after they had forged careers independently. An architect by training, Colin had moved into teaching and writing about architecture, while Hecht had worked with design office IDEO in America, Japan and London. They left these positions to create an alternative trajectory for a modern industrial design practice, taking an approach that was both optimistic and critical. Given that the Italian studio model of prolific individual designers, such as Richard Sapper and Franco Albini, holding intimate relationships with industry had long gone, this might have seemed a rather romantic idea. Industrial designers had become diluted in both their powers of persuasion and cultural position in the eyes of critics, museums and, indeed, manufacturers. Despite this, Industrial Facility found a way to open a new perspective, reaffirming the responsibilities that a designer should hold and communicate: namely, the acknowledgment that objects profoundly contribute to our everyday ease (and unease), our culture and identity and, at the same time, require the world's resources to do so. But to grasp the importance of Industrial Facility, one must first understand the context of their profession in its relation to the factories and companies that produce goods and services; and decide what we, as consumers, are provided with.

Setting

Industrially produced things are formed today from an unprecedented concentration of knowledge, attention and resources. Widespread digitized mechanization with low tooling costs allow manufacturers to respond swiftly to flickering market signals delivered to them by advanced analytic techniques. Markets once taken to be monolithic are revealed to be finely structured and unstable. Products, for their part, give form to the calculative processes that such conditions demand. They are, on the whole, opportunistic insertions, coded appeals to a precisely defined section of consumers forming micro-markets on a global scale. In this capacity, products act as singular points calibrated to hold market positions. Success depends upon a product's capacity to assert itself, on whatever grounds, over comparable products, until it must withdraw in the face of newer and better-equipped arrivals. At the level of our material culture, the consequence has been the creation of a listless and rebarbative object world. For, the flipside of the attention paid by manufacturers to the subtle relations between products within their narrow band of concern is a neglect of the environment formed by a combination of *all* the objects created on that basis. The profession of industrial design has, for the most part, adapted itself to these conditions. Designers find themselves caught between industry's inherent conservativism, born out of an extreme aversion to risk, and the fixed demand for renewal and innovation. The horizon of design activity is enclosed by the micro-markets defined by experts in other office suites. Its output, accordingly, tends towards a mildly speculative synthesis of proven demand, assembled in visual abbreviation on mood boards. The actual world – that is, the concrete space in which objects conceived by this process circulate and are put to use – doesn't get a look in. Inadvertently, the profession has retreated from the world of use to the abstractions of the market.

We need to hold these general conditions of practice in mind when considering the significance of Industrial Facility's work. For, although not unique in its resistance to the tendencies towards which the profession is prone, the studio is distinctive in its commitment to address the world though categories other than those of the market, while also maintaining an output that is emphatically industrial in character. The studio has created a great variety of products for an equally diverse set of clients. Among many other things, we can find a door handle, numerous seating designs, a telephone, a printer, a toilet brush, a complex system for organizing office desks and a knife sharpener. Surprisingly, given the number and diversity of clients and projects, this group of objects sits very well together. They seem almost to be a set. It's not that easy to put a finger on *why* this is. Formally, these objects do share a certain uncomplicated clarity. Reduced as they are, though, they are not slight; in fact, they have an unapologetic, even robust, presence. Yet, this common tendency doesn't quite amount to a style, it is more a question of character – something emphatic, an element of self-certainty. Ultimately, then, what Industrial Facility's various designs share (and what accounts for their familial appearance) is not first of all a question of form, but rather one of attitude, that is, a certain orientation towards the world, of understanding and engaging with it. It is the continuity of that project, maintained so consistently, and expressed at the level of form, which imparts to the studio's body of work such a high level of unity.

It may seem a little trivial to say of industrial designers that they are particularly concerned to take the world into account. After all, the primary purpose of industrial design in the production of commodities is just to help manufacturers to address a market. Design wouldn't get very far, particularly in the case of goods intended for mass production, if it didn't take the world into account. How else should design respond to an existing requirement, or anticipate a new one? Looked at in this way, industrial design would appear to do nothing but take the world into account. We might pause to ask just how, within such practices, the world actually appears. For, it turns out that what *is* addressed here is, in truth, only a narrow and peculiar abstraction of the world. It is a shabby version of the world of our ordinary day-to-day experience; one populated by goods and services, structured by a dense network of relations between market positions. In this hostile world of things, the worst aspects of our social world are foreshadowed in a tendency towards atomism and the elevation of competition to the status of a moral principle. In so far as contemporary industrial design views the world at all, it is through this myopic lens.

The radical aspect of Industrial Facility's practice derives from the studio's quite different relationship to the market. Although its output enjoys considerable commercial success, this is not as a consequence of the integration of design practice and market analysis. Rather, it follows from design in a mode that largely ignores established market trends and seeks instead to form independent criteria for the necessity of products, derived from a critical analysis of use. The difference between a given product's relationship to other products on the market to that of a user in the moment of use is as profound as it is simple. What follows is a description of some examples of Industrial Facility's work, selected on the basis of their form of engagement with the world, and divided accordingly into sections under the titles of Dependency, Borrowing and Commentary. The first two of these elaborate upon concepts used by the studio to describe its practice.

Dependency

A few years after Industrial Facility was founded, the studio published a book in two volumes, the second of which, *Product as Landscape* (2003), set out a concept that is fundamental to the studio's practice. The text reads as a statement of programmatic intent and, for all its mildness of tone, amounts to something approaching a manifesto, a literary genre seldom associated with design practice these days. It directs a radical reorientation in the way that products might be conceived in their relation both to each other and to their users. Under the heading 'Renaissance', a prevailing paradigm of user–object relations is identified, one that is taken to be problematic. In the history of ideas, the Renaissance is associated with the emergence of Humanism: the first incipient articulations of the modern notion that human beings, rather than the dogmatic authority of the church, should be considered the source of meaning and value. According to this anthropocentric model, products are seen to be designed on the assumption that they exist only for their users, a relationship so complete in its directness that it amounts to a denial of the existence of other objects. It is as though, in their individuated appeal to users, objects themselves had come to operate something like self-assertive, antisocial beings, convinced only of their own reality. At the level of environment, the first consequence of this paradigm is confusion – a cacophony of objects vying with each other to enter into a singular relation with users. Thus, a secondary consequence is a proliferation of redundancy, as functionality is greedily gathered up within individual devices in the hope of increasing opportunities for engagement. The Renaissance model of product individualism stems from a basic failure to distinguish between the marketing requirements of the shop floor and the concrete facts of use. The historical connotation of the term 'Renaissance' perhaps suggests that, however natural those relations may appear as we navigate from page to page to page, they are, in fact, our own outmoded production and, as such, are susceptible to change.

A corrective is given in the concept of 'Dependence'. Rather than considering objects to be a set of isolated and autonomous nodes, they might instead be thought of as necessarily 'co-dependent'.[1] Industrial Facility develop this idea through the figure of a landscape. A landscape is a thing with a distinct identity. Although coherent and formed of interlocking elements, this identity is not fixed. It is produced through a combination of parts that may change over time, or be removed and replaced by others. These parts may be varied in kind and need not tessellate neatly, nor add up to anything particularly attractive. Yet, whatever the result, in so far as a whole is formed, individual existence is necessarily subordinate to the totality, since it is only *together* that the parts form a whole. From the standpoint of the whole, then, the parts are mutually dependent, regardless of their other characteristics and differences. Transposed to the sphere of products and their uses, this way of looking at things requires a comprehensive re-evaluation of the status of objects. No longer designed exclusively on their own terms, objects might instead be thought of in a mode that is 'sensitive to proximity and to how things lie next to each other'.[2] If the inevitable contiguity of objects is taken into account, a kind of stripping back follows in recognition of the fact that they can be equipped 'with less, because they are supported by something much bigger'.[3]

It is important to distinguish this approach from asceticism. That objects today tend to be functionally and semiologically over-freighted will be apparent to anyone who has recently visited a department store or shopped on Amazon. But the point is not a reduction of individual items on the abstract principle that the idea of less itself should be taken as some kind of moral imperative (and the basis of self-regarding virtue). In the end, such an approach still involves the exaltation of individual objects as *humble*. Instead, a necessary paring back of objects is established within the context of the whole. For Industrial Facility, the drive towards reduction is not born out of an aesthetic of minimalism, but stems more from a functional holism. Although mildly stated by the studio, the implications of this approach for design are far-reaching. It counteracts the dominance of market relations over design, and draws attention instead to relations of practical utility, our common sphere of experience.

The concept of product as landscape is not an aesthetic formula. It directs a mode of attention, a standpoint on objects, governed by the principle of relation. Indeed, because objects are connected through relations that are always particular to them, the concept offers few – if any – rules to practice, beyond the requirement to consider those relations in the fullness of their particularity. Second Phone (2002) [pp.*200–2*], designed by Industrial Facility for the Japanese producer and retailer, Muji, offers one example of such a thinking through of relations, in this case those between a household's telephones. When the product was issued, the concrete situation of the domestic use of telephones was such that, while a household may have been equipped with a number of handsets, only one – perhaps positioned in a hallway or kitchen – was primary. This device and its many functions received frequent and comprehensive use. Meanwhile, the household's remaining phones received little use, if they were used at all. The bedside phone, for example, largely existed to remove the need to visit a phone ringing on a floor below. Its other possibilities were functionally redundant and visually disturbing. Second Phone provided a solution by thinning out unused function and visually compressing what remained. It is a low grey brick, just tall enough to have some presence as an object. The radiused, outward-facing edges make a minor concession to the hand that might enclose it, but hardly offer an invitation. Overall, the object has a certain visual neutrality; all of its 'interesting' functional elements – its mouthpiece, speaker and keys – having been positioned out of sight on the underside of the phone in its dormant position.

The design is thereby situated within a number of relations. The typical cradle of a telephone functions somewhat in the manner of a plinth for a piece of modernist sculpture. It is a dislocating device that mediates between the presented object and its physical setting, precisely so that the object is experienced on its own terms and not in relation to other things. Second Phone doesn't have a cradle. Its mouthpiece and switch are integrated, so that the telephone line is opened in the action of lifting the device, and closed when the object is set down. This means that in order to work at all, to avoid being permanently 'on', the phone explicitly depends upon the physical presence of something beyond its own limits – the provision by another object of some more or less horizontal surface. An acknowledgement of the object's particular station within the hierarchy of the household is expressed both in its material quality and in its functional poverty. Having been prepared within constraints tight enough to allow the phone to be sold cheaply, it feels light in the hand and sound quality is acceptable but not great. All of this is entirely correct and, given the particularity of its terms of use, marks the difference between a product designed with precision, even in the specification of 'poor' materials, and one that is merely shoddy. Given the thoroughgoing relationality of the design, it seems fitting that Second Phone is no longer in production, the home phone typology generally having been negated in the emergence of new communication technologies, specifically with the arrival of the smartphone. If a household still maintains a landline, what was once its primary phone is now relegated to communicating with (and screening) elderly relatives.

A similar criterion of correctness, established on the basis of a close examination of the concrete conditions under which a product is used, framed the design of the Picturemate photo printer (2006) [pp. *46–7, 52*], for Epson. The design follows from the insight that frequency of use varies greatly between printers in office and domestic settings. Whereas the office printer may be in constant use, the domestic printer is used only occasionally. This carries significant implications for form that led to the development of a new typology. Industrial Facility recognized that in the case of the domestic printer, usefulness is not identical with the effective performance of discrete tasks – it could also be something negative, a question of not doing something. In a domestic environment, a printer is maximally useful if, for the time that it isn't actually printing, it doesn't behave like a sports stadium organ. So, rather than offering an array of solicitously extended surfaces and blinking lights on constant display, the Picturemate presents an attractively subdued and solid form in its resting state. Only when printing is needed does it unfold to reveal the visual and formal cues that we tend to associate with printers. As an approach viewed from the side of design practice, the concept of product as landscape directs an interrogation of established product typology from an anthropological standpoint, and thus also entails an implicit rejection of the prevailing industry norms through which such products tend to be thought. The concrete context of use and definite relations between objects within such contexts provide a general framework for thought, from which problems of form are worked out.

Borrowing

Form is not only treated as a solution to problems of use. Many designs disclose a concern with use that is equally concrete but rather more particular, having less to do with calibrating appropriate levels of hierarchical presence and more to do with the particularity of the *experience* of use. In this instance, form is used as an instrument for activating attention within the act of use itself. As above, the means of this is not an exploitation of form as the occasion for expression, but rather as a means to create relations between objects. The impulse is, however, perhaps more intuitive than analytic: the studio refers to the operation as 'borrowing'.

The Passport Memo (2006) [pp. *59–61*] for Muji provides one such example. As the name suggests, the design consists of a notebook, to which the formal and material characteristics of a passport have been transferred. As a result, the object has the appearance of a passport – format, corner radii, page number and weight, binding – and the functionality of a notebook. At first glance, such a contraction may appear almost joke-like. However, the value of the operation becomes palpable in use, as the notebook is flexed in the hand, or slipped into an inside pocket. Embedded in the grain of these manipulations and gestures is a memory of the always deliberate siting of the passport, which distinguishes the placing of that vital article in a pocket or a drawer from other more ordinary acts of putting away. Small books for recording one's thoughts are not issued by the government but, depending on their contents, they may be as important as a legal document. The plain oddity of Passport Memo comments on this disjunction between personal and state paperwork, and the different kinds of 'journey' that each implies. It becomes effective as an object at the brink of conscious awareness, midway through an action, drawing attention to one thing by way of memory of another. Rather than distracting from the action at hand, this has the effect of heightening, by an incremental degree, an experience of use that might otherwise be merely automatic. Ultimately, the design's treatment of form is quite different to that initially suggested. A joke predicated on form is a kind of game played between meanings conventionally attached to certain combinations of material and shape. It offers an aristocratic mode of alienated enjoyment, an ironic pleasure in observing the operation of codes at a distance. Yet, experiences of this kind, even when they are pleasurable, are inherently dislocating. The affective aspect of Passport Memo locates the user firmly within the physical context of an action through a half-conscious reflection that enriches use, cutting against mechanical routine.

A more directly functional – if not indeed, functionalist – example is given in Door Tool (2001) [pp.*24–5*], an unrealized early concept design. The initial impulse came from an observation of the temporary door handles used on construction sites. Before the final fit, doors are opened using a simple lever with a square hole at its fulcrum. This is placed over the square-sectioned spindle projecting either side of the latch. None of the considerations usually brought to bear on the design of door handles apply in the case of this lever. It is an entirely functional means of getting at the latch located at the door's centre, in contrast to which most door handle designs appear almost as decorative apologies for the operation of doors. Even seemingly modern designs tend to focus their attention either on the ergonomics of the handle, treated in abstraction from its mechanical operation, or its industrial styling. In both cases, the handle conceals the underlying facts of door opening – that is, the interaction of lever and spindle that in turn engage the latch. The builder's temporary door handle belongs to a way of conceptualizing a structure prior to the practical coding of its spaces. Interconnected cellular enclosures are yet to become rooms; doors are inconstant barriers. Door Tool retains something of this reduced, pre-social functionalism. It is a cast- and machined-aluminium lever that expresses its junction with the square peg, which is exposed at its centre. As the lever is pressed, the peg is seen to rotate. If only for a moment, the door opener recognizes itself as an *operator*. Both Passport Memo and Door Tool offer gentle interruptions to habitual patterns of use through a displacement between object types. By careful attention to the use of one, the use of the other is heightened. This operation is subtle, perhaps even poetic, its success depending more upon strength of intuition and courage than on strategy. It provides users with the opportunity for a renewed connection to objects in use. More generally, the approach both discloses and prompts an awareness of a necessary partnership with objects in our interactions with the world. This is fundamentally quite a different mode of relation to the use of objects as communicative tokens. 'Good design,' as they say, 'is radical only because of its uncompromised essence – its form and communication are relevant to the life of the product and not only to the shop.'[4]

Commentary

The two tendencies that I have referred to as Dependency and Borrowing involve a conception of objects that pushes against the conventional patterns and assumptions within which product typologies become fixed. In different registers, they draw our attention back to the world. The first considers what is required of objects if they are rethought according to the totality of functional relations in which they are embedded. By considering the actuality of these relations, in place of the abstract requirements of marketing, new typologies appear and contribute to a rebalancing at the level of the whole: the world is thereby revealed more clearly by scraping off some of the redundant clutter with which it has become encrusted. Use, in this case, is approached as a distributive problem that finds modest adjustment at the level of individual objects. Borrowing also breaks with assumptions governing the connection of form to use. But the connections that it opens between objects are oblique rather than systematic. The heightening of experience that results also refocuses attention on our concrete interactions with the world. Although Industrial Facility have produced a number of designs that seem to fall cleanly into these categories, some combine them to varying degrees. Others suggest another, less directly engaged, mode of relation to the object world that, for want of a term supplied by the studio, I am calling 'Commentary'. As with the other approaches, designs within this group are predicated upon an observation concerning, for instance, the way that shared use of public space is negotiated. Unlike Dependency or Borrowing, the terms of use are not contested but are simply revealed in a way that assists use according to its existing pattern.

One such design is Table, Bench, Chair (TBC) (2009) [pp. *92–5*], a seating system for the English company, Established & Sons. The design is intended for use in hallways, lobby areas and waiting rooms. Transitional spaces like these are curious, as their purpose is simply to mediate between other spaces. Occupying them implies subjection to administrative processes that are voluntary to the extent that they are accepted, but rarely chosen. A certain ambiguity in the status of these spaces is duplicated in the uncertain outcome of the processes that they serve. This unsettled state registers in the design's abbreviated name, TBC being bureaucratic shorthand for arrangements yet 'to be confirmed'. There is some oddity about the character of TBC itself, which seems to acknowledge the strangeness of its intended environments, and perhaps the awkwardness of social transactions that occur within them. The horizontal surface of an unpainted oak plank provides a functional plane for sitting and resting objects. A framework of steam-bent beech dowel supports the horizontal plane and divides its upper surface into zones of varying use. Table, Bench, Chair is a modular system and, as such, is highly flexible, providing for a number of combinations according to need: the addition of a frame to the surface directly above a set of table legs creates a chair, defining a sheltered zone for its occupant; the absence of such a frame implies a communal bench; the extension of the horizontal surface beyond the frame of the chair makes a side table; a love seat is formed by reversing the orientation of one of the pair of seating frames. In its functional aspect, the system is rigorous and highly rational. Visually, on the other hand, it is strange to an almost unsettling extent. For, at the same time as it serves as the indispensable basis of the system, the continuous horizontal surface also appears to slice disruptively across it. Figuratively, the seating system projects the image of a violent transaction between a delicate seating group and a large plank of wood. In this way, TBC responds to its particular environment of use with a double commentary: on the one hand, offering demarcation of an uncertain territory, and on the other, acknowledging the social peculiarity of a space in which such division might be welcome. All of this stems from a generous anthropology, attentive in its observance of the particularity of customary uses.

Run (2016) [pp. *212–15*], for Emeco, is a seating, table and shelf system that also responds to a particular social need for the definition of personal space in a situation of common use. However, this more sober and utilitarian design is intended for areas such as the canteen or outdoor seating which, unlike the waiting room or lobby, are inherently social. The design problem in this case was how to maintain the possibility of defining personal space without discouraging sociability. A balance between these requirements was made that prioritizes the social aspect. Run is derived from the picnic table and bench, typologies associated with collective use. These open and longitudinally extended forms imply non-hierarchical gathering, a spontaneous social convergence from many directions. Yet, the open-plank construction of the tabletop, which signals conviviality, also responds to the need for privacy. Gaps between planks serve as informal boundaries, marking the limit beyond which the extension of a laptop from one side of the table may be experienced as an intrusion on the other. The divisions formed in the gaps between planks thereby imply the optional basis for a certain consideration of use, one that may be rigidly observed or entirely ignored. A range of accessories are provided that, using plugs located in the gaps between planks, can be used to impose a more architectural division of space where there may be greater territorial need, such as in a library workspace. Two extrusions – one for the leg, the other for the plank – serve as basic system elements, out of which shelving, table and benches are arranged. In a more figural mode, this systematic coordination of elements also conveys a social significance of collective endeavour, suggesting a functional whole produced through a collaborative combination of equal parts. If the product atomism, critiqued by Industrial Facility through the concept of landscape, mirrors the general form of social life, then the relation of system elements from which Run's tables, seating and storage are made might suggest an alternative arrangement, one based on an equality of association, unity and, perhaps, solidarity.

History

As a set of objects, the consistency of the studio's output suggests the result of a project. Such an even continuity of attention is far from the norm within contemporary industrial design practice. Indeed, taken as whole, there is something seemingly historical about the degree of coherence in Industrial Facility's body of work. It reveals a focus of purpose and continuity of activity that one might associate more readily with the post-war period of heroic Modernism. The conditions that sustained projects such as the Braun Design programme of the 1960s no longer hold, however. So, while on the face of it Industrial Facility's reduced forms and the German company's adventure in functionalist programming may appear to hint at some underlying commonality, in reality the two could not be more different. Braun Design grew out of a reconstructive effort to remake the world in a rational and more human form following its near total destruction by irrational forces. This was made possible by the combination of an economic boom and the fact that the science of market analysis had yet to significantly impinge upon the freedom of designers. Industrial Facility practice under diametrically opposed conditions, however, and their project is formed accordingly. Their aim is not modernist in the utopian mode of world-making by industrial means. It is rather more corrective than it is constructive, for it addresses the alienated and abstracted world of capitalist development, which first stifled and then overwhelmed the utopian projects of the last century. Under these constrained conditions, Industrial Facility's approach is necessarily more tactical than strategic. Its aim is to find routes through industrial design by which users might push against the effects of those processes, in order to recover a more concrete level of contact with the world.

The consistency of approach evident at a formal level in the studio's work is remarkable because it seems to suggest a freedom of practice that is generally absent from the profession. In truth, it results from an extraordinary level of care on the part of the studio in considering who to work with, when and on what terms. The clients that Industrial Facility work with are themselves atypical, in the most obvious sense that they are mass-manufacturing companies who are yet willing to accept designers whose products succeed in the market precisely on the basis of their rejection of the kind of standardized, precautionary steps generally taken to avoid failure. These are companies, in other words, that take design seriously as a process, out of which meaningful products might be developed rather than decorated. And they share with the studio a certain generosity with regards to the customer – a belief that, in the absence of other inducements and seductions, people (or enough of them) will get it. The relationships that the studio enters into, with such clients as Muji, Wästberg, Mattiazzi and Herman Miller, to name but a few, tend to be longstanding, discursive and analytic. On the client's side, this is rooted in openness to the possibility of a successful result, the particularities of which could not have been predicted at the start. On the studio's side, a condition of that success is the attention paid both to the requirements of the client and to the possibilities that their particular expertise as manufacturers affords. Thus, for Muji, there is a series of objects, many produced in plastic and manufactured according to tight constraints, each deliberately self-effacing but contributing to the formation of a programme of products of unprecedented scope and coherence. Industrial Facility's various contributions to this programme are born out of years of discussions and workshops that have established an unusually profound level of accord between the studio and its client.

Although Industrial Facility are certainly busier today than when the practice was founded, the studio is not significantly larger: it is presently staffed purely by its two directors and three assistants. This is as telling as it is unusual. Despite the success of their products on the market, which might have given the studio a basis for expansion, Colin and Hecht have chosen to remain directly involved in all of the projects that they take on. The reason for this is simply that design, for them, represents a total engagement with a problem; a process of working through to a solution that cannot be broken up, distributed or merely supervised. At this level of attention, there is a direct correspondence between the mode and quality of engagement required to arrive at a satisfactory solution, and the forms of experience that those products seek to provide for their users.

An inevitable consequence of Industrial Facility's success has been a weak kind of formal influence. One now frequently comes across products that have something of the Industrial Facility look – a certain radius, a particular density, a determined avoidance of decoration, or an awkwardness that seems to indicate some undefined functional gain. Even design schools have set projects for their students to design 'in the manner of Industrial Facility'. As they wryly note, there is a particular perversity in what one has to assume are the frequent appearances of the studio's work on design consultants' mood boards. It would be a shame if superficial styling were the extent of the studio's influence. But the useful model presented by the studio, particularly for younger designers, is less formal than existential. If Industrial Facility are to be emulated in a meaningful sense (and there can be no doubt that they should be), this would require a deliberate commitment to industrial design, in contradiction with its present tendencies, as a form of engagement with the world.

1 Sam Hecht and Kim Colin, *Product as Landscape*, Vol. 2 (London, 2005), 7.
2 *Ibid.*, 6.
3 *Ibid.*
4 The Design Museum, *Designer Maker User* (London, 2016), 82.

Selected Projects and Interviews Part One

Door Tool pp. 117, 257

Twinstopper pp. 117, 268

Wireframe pp. 118, 271

City in a Bag pp. 119, 259

Magnetic Tape p. 272

Bruce Mau with Sam Grawe

Everything about Bruce Mau just seems a bit bigger than the rest of us. He's a big man with a big smile. He dreams big dreams and wrestles with big ideas. He collaborates with big names and produces work for big clients. He makes really big books with titles like *Massive Change* (2004). Because his work defies pigeonholing – in medium, method, content and concept – 'big' may even serve as an apt descriptor for his omnivorous design practice.

In 2012, Sam, Kim and Bruce were ensconced in a speculative project operating under the code name 'Together'. From their offices in London and Chicago, and deep within the back hallways of Herman Miller's Design Yard, they developed a programme that reimagined furniture production, sales, delivery and assembly in a way that was entirely different to what had gone before. 'Bruce was able to make our inventions and designs understandable to the common person,' Sam recalls, 'and at the same time, to turn the work into culture. He made the whole project, which was relatively simple, into a kind of platform for change.' Ultimately, it was more change than the company could handle and, for various reasons, the project never happened. While Sam and Kim were devastated, Bruce moved on quickly. 'I think he'd been through situations like this before,' says Sam. 'For us, it took a bit longer to come to terms with the fact that ideas are not always enough to make a project happen. It was perhaps too ahead of its time.'

My relationship with Sam and Kim goes back to the early Noughties – to my days as the editor-in-chief of *Dwell*. From the outset, their thoughtfulness as designers – both aesthetically and intellectually – registered deeply with my own concerns. But, ultimately, it was the subject of music, not design, that sparked our friendship. When we met again in London, I gave them a copy of my first album, *Colors of the Sun*. The opinionated journalist that they thought they knew was in fact a fellow creative spirit, and this shift in perception ignited a mutual admiration that has come to define both our friendship and our professional relationship. I'm always impressed by their forthrightness, their attention to detail, and their unfaltering desire to make things better. So it was with pleasure that I made the short trip around the horn of Lake Michigan on a blustery autumn day to sit down with Mau in his Chicago studio to talk about our friends Sam and Kim.

Sam Grawe: Tell me about the first time you met Sam and Kim.

Bruce Mau: I have a very poor recollection of that kind of thing, but I have a very good recollection of their impact. I think that the talent they have is so formidable – they have an ability to make everything that they touch beautiful. And not only can they do it themselves – that's one kind of talent – but they can also instruct other people on how to do it and really go into detail on the very subtle things that are actually going to make a difference.

That's why it was such an absolute delight to be on the Herman Miller Design Advisory Council with them. When they spoke about something, I was fascinated to see how they saw it. To witness what they were seeing. They would identify the one thing that didn't work, or would show how subtly changing the radius of a curve could make a design sing. They both have that ability. Very few people I've ever worked with, or met, have quite that combination of talent.

Could you take us back to the founding of the Herman Miller Design Advisory Council (DAC)?

The phone rang in 2008, and on the other end was Don Goeman. He explained that Herman Miller wanted to put together a group of people to work with the CEO, Brian Walker, and the senior executives to help them apply design thinking to their whole business. That resonated with me because I think it's how I would have approached it if I were going to suggest how to do it – not only to think about the design of a product but to also think about the design of the whole enterprise.

Herman Miller did a great job of selecting people who could contribute in different ways, and understood how to be a part of this sort of group. I guess they had tried it before with disastrous results – like an actual fist-fight – and the moment Don told me that story it sounded familiar. There are certain kinds of designers for whom their world view is their world – and that's what they do. And then there are other kinds like Sam and Kim, who I had never met before. They have the capacity to empathize and travel outside of their own domain to understand what other people are struggling with, and who are in fact able to help understand and articulate what the real problem is in the first place, all before any solutions are developed. That's what Sam and Kim do well.

What effect ultimately do you think that group had?

It was pretty transformational, I think. I recall a discussion at one point about how the people who had presented to us felt like they were standing before some kind of tribunal. People would be asked to prepare their work and present it to us, and we started to get these reports that people thought it was traumatic, because if they were going to get a negative critique, they got it in front of all their bosses too. So we ended up having to say to the company, 'Look, we're not here to judge you. We're committed to helping you to advance your work.' And after that, it started to work out.

I remember one of the discussions that was most profound to me, which was around a project for an inexpensive chair that people almost seemed a little bit embarrassed about. We basically said, 'Look, any bonehead can design a $10,000 chair – but what about a $100 chair that has the quality of Herman Miller? That's a beast of a project! Why aren't you working with the best designers in the world on that – getting your best people assigned with that objective?' It was that kind of thinking that I think helped to shape a new approach and strategy.

So it seems that the DAC played the role it was supposed to play – *and* it ended up bringing you together with Sam and Kim.

Sam, Kim, my wife Bisi and I all just had such a simpatico kind of feeling when we spent time together. It led to us talking about things that we saw were opportunities for us to collaborate further on. We'd say, 'Wow, this would be so cool if we could do this.' So, that led to the prospect of collaboration some day.

So that's how the Together project initially came about? Sam said that he became obsessed with the idea – that he even had dreams about ways to make it work.

It was pretty all-consuming. It was a beautiful concept: we had this idea that you could put together a piece of furniture in a few seconds without any tools. And we decided that the way to sell Herman Miller on the idea was to

physically make the product faster than we could actually describe the assembly of it in words. So I was on one side of the room walking through the instructions, while Sam and Kim were actually doing it in real time on the other side. Brian and the executives couldn't believe what was happening right in front of their eyes. It really was a stunning piece of work. We didn't get the chance to put it all the way through an industrial manufacturing process, but it was a really magical moment in my career. I mean it's tragic that we didn't ... let's say that we haven't yet produced it.

So there's still light at the end of the tunnel?

I have to say, I am a really long-term player. I never give up. I'm still working on projects that I started thirty years ago that haven't been produced yet. Sometimes you're just doing stuff too early. The market, the money, the world: things just aren't there yet. And it sometimes takes decades for the world to make a place for it. I just do that naturally: that's just the way I am.

One of the things that interests me about Sam and Kim is the fact that they are partners in life and in design. Can you give us a little insight in to how you regard their relationship? Do you believe that it produces results that neither of them would come to as individuals?

It's interesting because I've worked with a few people who worked that way – the architect Frank Gehry and his wife Berta, for instance. When you're working with someone who's your life partner as well as your business partner, it's a very helpful voice. I think that it allows you to see things in different ways. Bisi, my partner in life and business, has incredible insight into my work in ways that I don't. She sees possibilities that I don't see, and she understands context in ways that I miss. She has different passions and gets inspired by different things. She can see the field from a different vantage, and I think that's pretty powerful when you put it together. I've seen it work that way with Frank and Berta, and I've seen it work with Sam and Kim.

I also think that, in their case, the hybrid of American and British culture is interesting, because Kim brings a different energy and perspective and engagement that the British are just never going to do. And similarly, Sam has a kind of intimacy and thoughtfulness that is powerful. They're both clearly talented and capable in their own ways. But together, they're even more powerful and capable.

From what I've seen, they share a lot of the same instincts, but they seem to bring wildly different, and complimentary, skill sets to the table.

Yeah. The same is true with Bisi and I, and I'd say so for Frank and Berta as well. I've seen Berta say to Frank, 'Frank, that's a terrible idea. Don't do that.' That honesty is important. When you love somebody, your commitment to their well-being is really deep and explicit. And so, if you see them doing something that is ultimately not in their best interests, you're willing to actually tell them that!

For Sam and Kim, I think that operates at a really high level. There's so much love and commitment that honesty is inevitable. Like, it's got to be there.

That certainly echoes my own experience of working with Sam and Kim. In any situation – whether it's offering criticism, praise, displeasure or gratitude – there is a sense of resolute honesty.

I think that's what made them really good on the Design Advisory and that's why I love them. They're great. They're great designers. And great designers have integrity. A real designer tells the truth, you know? Because, otherwise, you're working on the wrong problem.

For me, empathy and integrity are the key values of design practice, and so I have to tell the truth – like it or not. Sometimes, it's really awful to tell it; someone's put a lot of work into something and they think it's great – but I have to tell them how I see it. It's not necessarily the universal truth but at least it's my truth.

If you had to describe Industrial Facility to someone who had no experience of their work, and you couldn't show them any pictures, what would you say?

Intelligence; clarity; simplicity; beauty; and really in that order. It always starts with intelligence. They start with a really smart insight, so that they've got something to do that is a smart thing to do. And then they get clarity on it – they hone it down to exactly what it needs to be. They make it simple, which is the hardest thing to do. And then, they deliver it with beauty. I've never seen a bad thing they've done. There are some things that I love more than others, but each thing is still beautiful!

What would you imagine their lasting contribution to the field of design might be?

They push that intelligence all the way through. It always seems to me that the results are emblematic of what humanity is really capable of. And if you think about that as a legacy, I hope that they keep going for a long, long time because the more that they can demonstrate it, the better off the world is going to be.

I truly believe that fear has become the dominant idea of our time – you see it just about everywhere, every day. And so when we're living in this era of fear and you have designers like Sam and Kim who are fearless in their thinking, it inspires everyone around them. The work inspires people who come in contact with it, because it says, 'We're still here, we've still got lots to do, and we can do it!' That's a profound message to the world, and I think that's ultimately what great design is about. Against these tough problems, we're still going to create. We're still going to find ways of being smart. And, for me, they just embody that ethos perfectly.

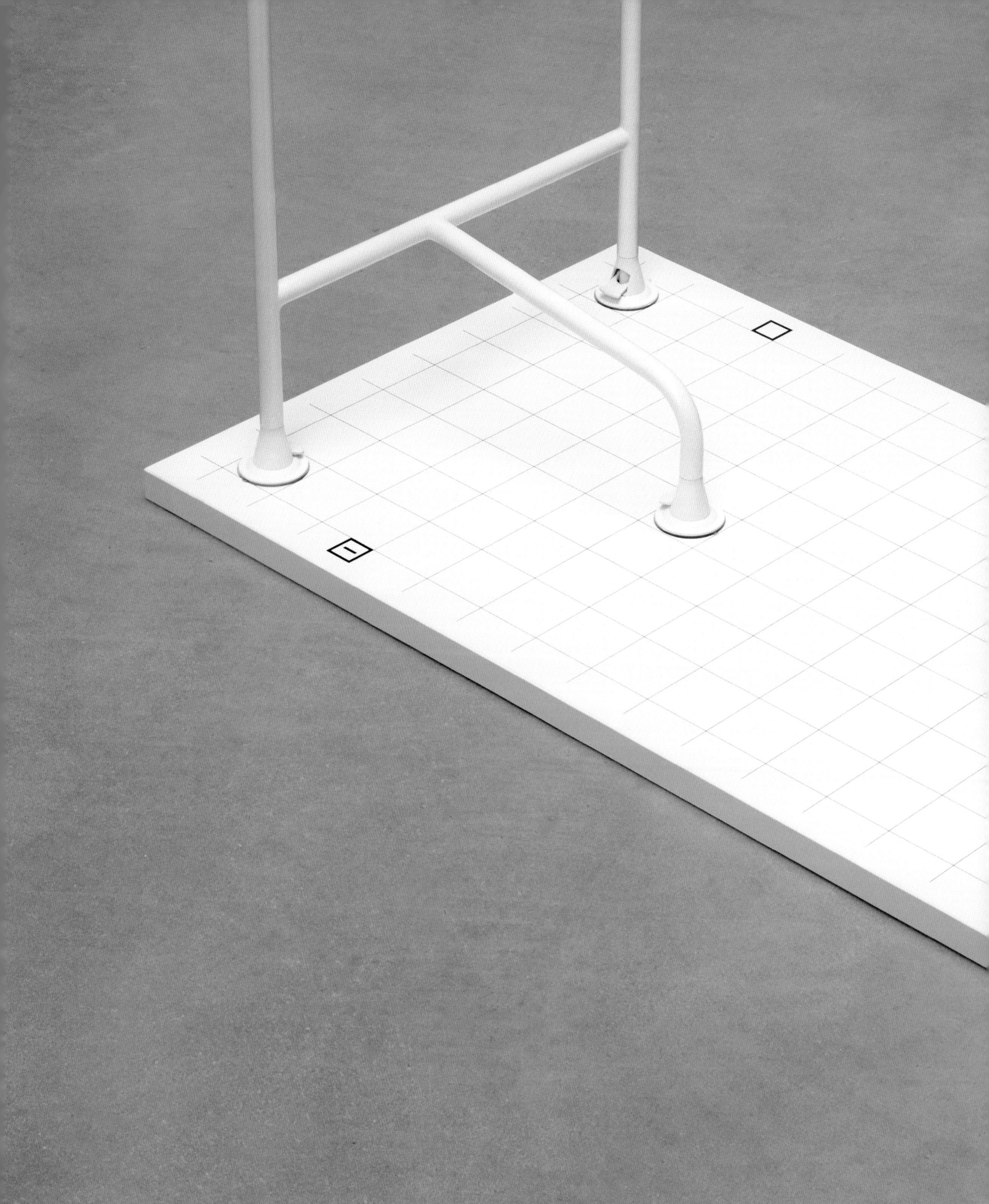

Together pp. 120, 274

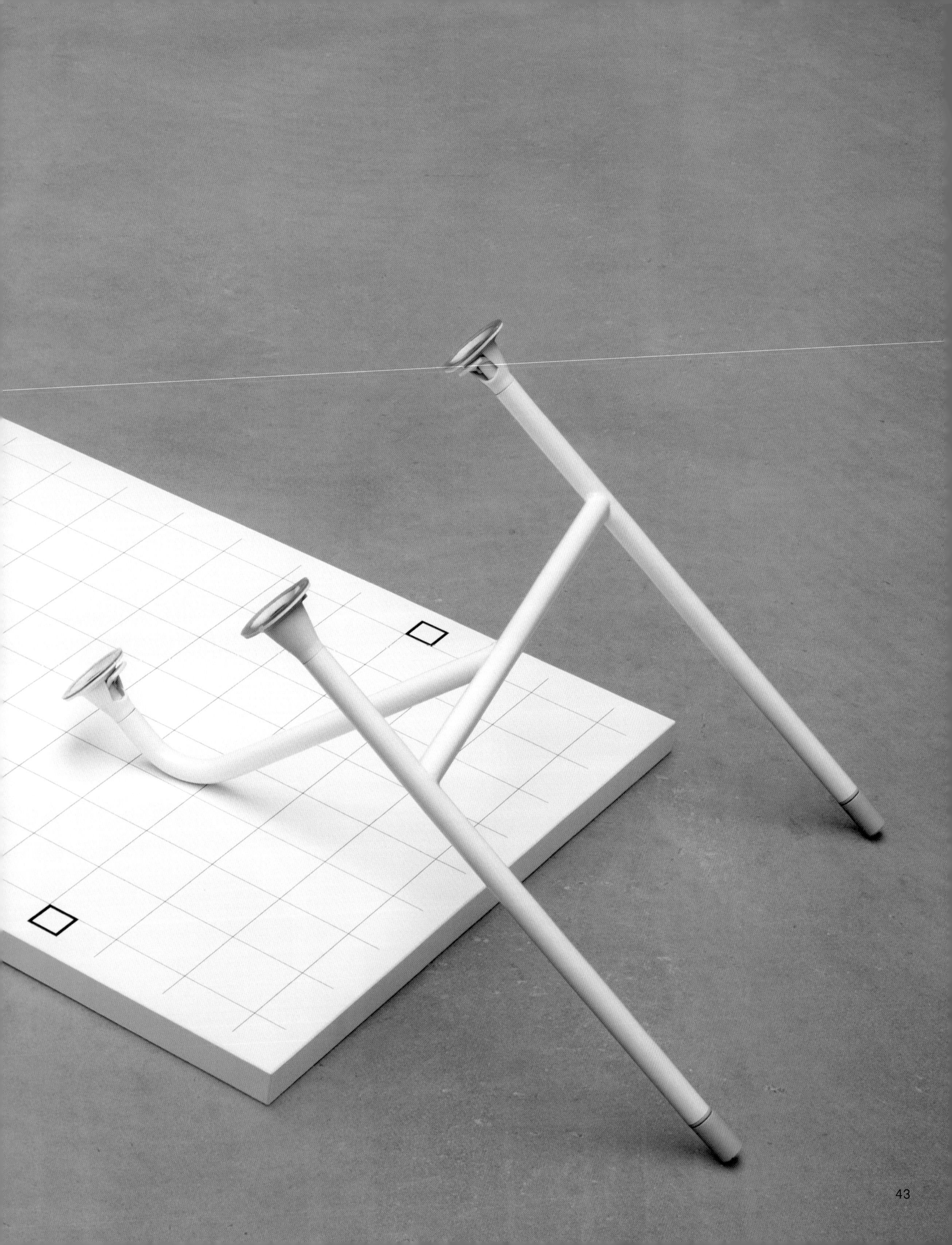

Picturemate pp. 121, 262

EPSON

28
26

29

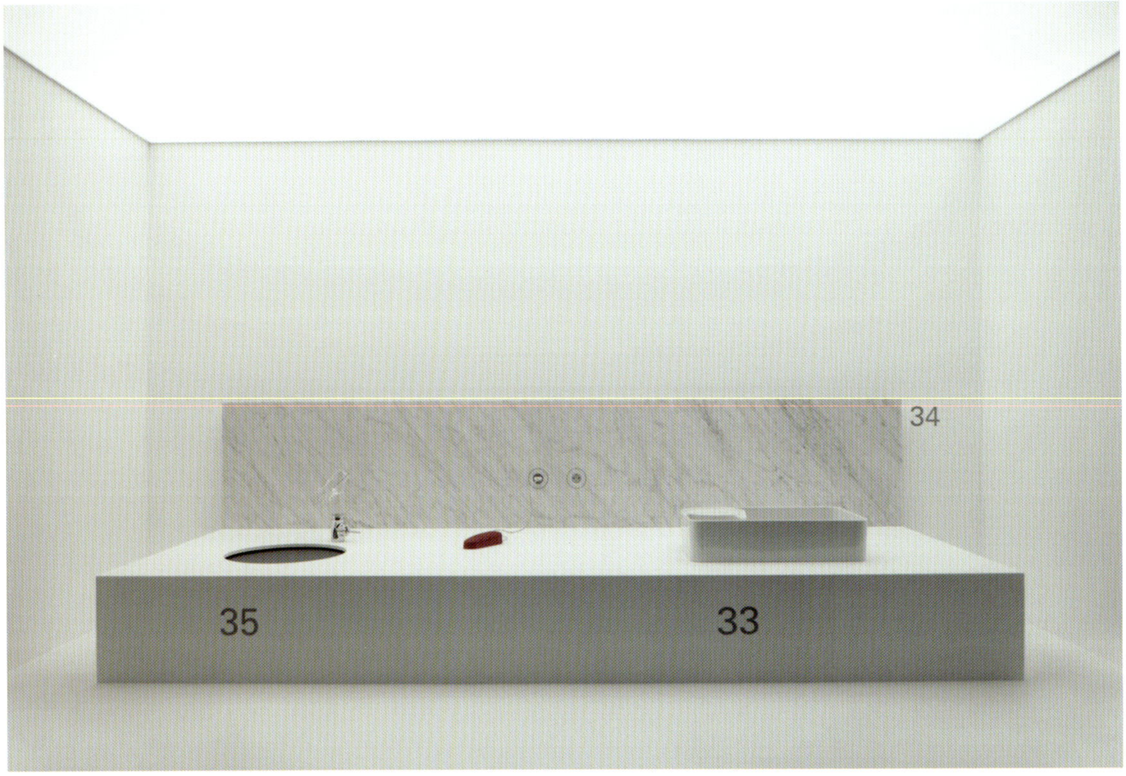
34
35
33

EPSON

Circumference pp. 123, 265

Fan pp. 124, 262

Cycle Shirt pp. 125, 270

Pentagon pp. 125, 270

Branca chair pp. 127, 269

Branca table pp. 127, 269

Naoto Fukasawa

Landscape Camera pp. 129, 257

Naoto Fukasawa with Eizo Okada

As you walk deeper and deeper into the avenues of Tokyo's Aoyama district, past the luxury boutiques, you eventually come to a quiet residential neighbourhood. That's where the designer Naoto Fukasawa has his office. I arrived one Friday evening after Sam Hecht suggested that I talk with him. I followed his staff into a spacious one-room loft studio that serves as his office and found him sitting at a big table, writing some kind of note with his back towards me. He stood up, greeted me with a friendly smile despite looking a bit tired, and explained that he had just returned the day before from a short trip to Europe.

Eizo Okada: Naoto-san, please tell me about the first time that you met Sam.

Naoto Fukasawa: I first met him in San Francisco when he had just joined the design office of IDEO. I think it was soon after he had graduated from the Royal College of Art. I immediately felt that he had a different sensibility compared with other designers I knew. At that time in the early Nineties, design strategy for innovation was the main focus in IDEO, and I too had honed my skills in shape creation, and made use of those skills in many projects. But Sam didn't seem to think that design was only about creating shapes. Rather, he put weight on what he called 'the found object' – looking for inspiration not in form but from existing objects.

Also, at this time, design started to change in a big way. I felt that his way of thinking and way of looking at things was very fresh, very different. And it seems that he thought that people like me who were skillful at creating and developing forms for manufactured goods were innovative and interesting.

Sam has said that he learned 'his trade', as he puts it, from you. What kinds of things did you lead him to?

He did? Really? I never felt that he followed too much of what I had suggested to him. I kept telling him that he should become a graphic designer. He had such strong editorial skills and was good at accurately expressing sensory perceptions, which is not easy. He was also good at putting things into words and describing them succinctly. As we worked together, I would laughingly say something like, 'Maybe you'd better change mediums, as your talent for description is in something other than just creating form!' Still, he kept on creating things, applying a different approach to industrial design, and didn't give up. I learned a lot from his 'ready-made' idea of design, which was a much softer and more humble way of design.

So, would you say that he was demonstrating an almost un-emotional design method, in a design world that was dominated by an overwhelming desire to create shapes?

His 'found object' style of design, where existing objects are appropriated into new ones, cannot be accepted just as is, or at face value. You still have to set the form, and you cannot avoid this. In the end, you have to finish the design down to the smallest detail. I, myself, was strictly aiming to design things that could fit into daily life, whereas his designs looked untouched, almost primitive. But that was the way he was then.

If you think of sushi as individual ideas, then it's really a matter of how much soy sauce you put on it, to accentuate its flavour, isn't it? You know how much soy sauce to put on because you know the flavour of each fish. Similarly, you can't lie to yourself about how well you shape an idea of a design, because you know its true function. I would often discuss with him that it's important to consider this because even if your idea is very pure, if you get the form wrong you will come out with something very different. We would critique each other's work constantly, with me saying something like, 'That's your (Sam's) way of creating shapes that don't seem to follow directly from the idea itself.'

You then invited him to join you in Tokyo. I have often felt that Sam's design sense is very compatible with Japanese culture. How did Sam cope with living in Japan?

I had taken over the Japanese office, and as I needed someone to work together with me, I knew he would be good and so I invited him. So Sam came to Japan and he fitted in immediately. But, as he's someone who definitely has his own lifestyle, he's different from the foreigners who love everything about Japan and become obsessive – he didn't come to try to be Japanese or anything like that. He's naturally sensitive to where he is, but he doesn't change himself. He's not the type of person who will try everything for the sake of trying it: he's careful, and because of that his work is very clean and precise.

Was the work you were doing at the time basically consulting for businesses?

We had a lot of projects and within that were workshops with different clients. I'm not the type of designer that really gets into workshops, but when Sam started doing them he saw a chance to introduce our new approach, and I started to see the potential in them too.

Commercial design at that time had so many restrictions. There was pressure from the marketing side and there were also many company rules along with the complexity of the production line to deal with. I felt that a company might get very confused if you tried to bring in a very different way of design. So that's why we did these workshops. Nothing like that was being done in the US either, and when I gave a report to IDEO headquarters, they thought that it was an interesting idea and gave us the thumbs up to continue.

Your designs progressively became simpler after you moved to Japan from the US.

Yes, I stopped putting 'meaning' or 'semantics' into the creation of shapes themselves. There was a time when meaning had to be put into every form, like computer monitors for example, which were big and didn't have much meaning to their shape. The manufacturers absolutely wanted some meaning to be put into the plastic housing because it was so big. And so the designer who possessed the greatest skill in creating shapes was the most valued – at the top of the mountain. Since meeting Sam, I've come to believe that rather than adding meaning to the shape, it's better to concentrate on how to simplify the expression and get the proper proportion. It's perhaps a more truthful expression.

Speaking of simple, I understand that Sam really likes Muji. What does Muji mean to you?

As I'm a director at Muji, and deeply involved with the company, I think that my feelings towards it are slightly different to his. Sam has been involved for a long time with Muji, and his products are selling well. His own style of living is very much in line with the Muji philosophy, and as a designer that understands Muji exceptionally well, I think he's someone that can live the Muji lifestyle.

The value or appeal of Muji is different from that of high-tech industrial products or high-end furniture, isn't it? The products seem to blend into the background of daily life.

I think it's often accepted that Muji products will do reasonably well in life, and that you can be satisfied with them. They give that feeling of being just right. A person can feel satisfied that even if they wanted something better, it wouldn't be that much different from what they can get at Muji. If you don't want anything better than that, then Muji is perfectly good. There is, however, a deeper aspect to Muji, which is that the products are very purposefully designed, based on what we call 'micro-consideration' or, in other words, a detailed examination of how we live. We make a lot of things that other manufacturers wouldn't even bother to consider, like our travel washboard for example. Sam likes those kinds of things and has amassed quite a collection.

Is that because you sense the wisdom behind these small everyday items, perhaps?

Yes, Jasper Morrison likes them too, those kind of folksy items. I like them as well, but I'm interested in creating products that combine a wider variety of elements. I'm good at creating shapes that people can naturally embrace. In my thinking, you can't create such shapes solely on rationale. Konstantin Grcic says that I'm the kind of designer that 'turns a frog into a prince'. I guess that it's an analogy for how I can take whatever you bring me and somehow try my hardest to make it even better. I don't know if I can say that Sam also considers this rationale to be as important to him. At any rate, he's one of the most intelligent people that I have ever met. He's clever, and there are times when I've thought that I'd like him to design even more intuitively because he's so smart. He's so logical in his approach.

Coffee Maker pp. 129, 261

ヒーター
OFF
ON

IF4000 knife rack pp. 130, 261

Chantry Modern pp. 131, 257

IF4000 knives pp. 131, 259

Bottle pp. 132, 274

Wine Opener pp. 133, 265

Radice pp. 133, 273

Dalston pp. 134, 275

Alumi pp. 135, 272

Paper Cup pp. 135, 259

Table, Bench, Chair pp. 136, 266

Durrell Bishop

Durrell Bishop with Duncan Riches

Industrial Facility has applied their trade to so many varied industries that they are probably the epitome of the multidisciplinary design office. I know them for their furniture designs, as they have created some of my favourite chairs and lights; but, equally, I store my data on their hard drives, make calls on their phone, visit their exhibitions and have enjoyed reading their books. Of course, to achieve this much variety in their work they have relied on a wide network of friends who have influenced them. Durrell Bishop – himself a talented character who has variously been a photographer, product designer, interaction designer, teacher and entrepreneur – is one of these.

Duncan Riches: When did you first encounter Sam and Kim?

Durrell Bishop: Sam and I had finished our Masters at the Royal College of Art in the early Nineties – his was in Industrial Design, while mine was in Computer-Related Design. A mutual friend, Winfried Scheuer, had been teaching Sam and said I should meet him, and that his work was lovely, very different. I remember that he'd investigated a hotel room, designing everything in it and that there was a particularly nice telephone. A few years later, when I was working at Interval Research and living in San Francisco, Sam turned up at the American design company, IDEO, after Bill Moggridge had pulled him in to do a project that first took him to Harvard University in Boston. I ended up giving him my spare room. So, for a while we were both living together in San Francisco. Sam was amazing because each evening he would bring back products that he'd bought in the street. I didn't realize it at the time, but this was the start of his collection previously called 'Under a Fiver', or 'Usefulness in Small Things', as he now calls it. I vividly remember him once bringing back a self-inflating envelope – you could put things into it, undo a tab on the outside, and when air entered it, it inflated around the package instantly. He had a fascination for collecting unusual inventions and odd products, unusual ways in which people do things. So, that's when we got to know each other. When he started his office in London with Kim, she was still teaching at the Architectural Association. It was an odd coupling, as she had no product design background. Yet, like me, she was intrigued by Sam's 'Under a Fiver' project, so I could see that they were perfect together – the odd couple.

Then later, you taught together?

I've taught on and off at the Royal College of Art for a long time and on many courses. Sam was asked by Ron Arad to put together a platform, and suggested that I join him alongside André Klauser who is a furniture designer. We based the platform on people looking at ordinary things around them, products that had not really been touched by good design; the things that you have to live with out of need but that you really don't like, like fire alarms or English plugs. Sam was the perfect person to teach with to get students to look at the function of simple things and to develop beauty through ideas.

Did you perform different roles on the course?

Yes. Something happened to me when I started to play with computers. I became interested in how people perceive object structures both code-wise

and physically, and I started seeing a different side to design – the relationship between a product and a service. If you think about it, products are simply descriptions of potential. They allow you to see potential in domesticity, in work and in society. Without the physical product, it's hard for people to see potential. People can't sit down without a chair: you need to know the chair is there to know that you can sit. You could say that a chair is a description of an action. This train of thought sat perfectly when students looked at Sam and Kim's work too. They raised their game in terms of thinking around the qualities and the physicality of the object. That's important. There aren't many designers out there who don't compromise the quality and point of view of their work like Industrial Facility, and in product design there are even fewer.

I am interested in where non-design-related activities come into play on such a course. How does art or humour enter into the discourse? I understand that you got artists involved on the platform.

We invited artists to come and talk to the design students. One of the artists, Richard Wentworth, was fantastic. We set up a project called 'A walk with Richard', and I remember that we all arranged to meet in Colindale, North London – a place that no one would go unless you lived there. We just wandered through the streets talking about the nature of the things that surrounded us. We discussed the correlation between the wealth of societies and the size of paving stones (you only get very large paving stones when people can afford the labour to move them). His anecdotes and thoughts about the physical nature of the world were something that we wanted to expose our students to. Product design should be born from an understanding of the physical realities of our world. Another artist that Sam brought in was Martin Creed, who did a whole other kind of talk about people and objects; and rather than have the students design something, they had to write down a description of operating an existing product in the manner of a set of instructions. It was wonderful.

How do you think Industrial Facility manage to convince their clients to do something different from the norm?

It seems pretty obvious that they have their own way of working with clients, which allows them to start in a place that most product designers can't start from. If you are working with a client properly, and you're talking about having time to actually understand what the product does, then you can design it differently. If you change the rules of it, then you start getting different products out, and that's what they do.

Their approach seems quite conversational; it's like they start by initiating a conversation with the client, and clearly some clients want to have that conversation more than others. When I interviewed Sam and Kim in 2009, I was intrigued by how they managed to get clients to accept the fact that they would likely challenge the design brief itself.

Well, if you do good work, you start getting the clients as well. You get the clients who want the quality of work that you produce. It's extraordinary to me that I've come to know Herman Miller through Industrial Facility mostly, and it's obvious that it's Kim who has made the depth of the relationship expose a layer that people maybe thought had disappeared. They're interesting as a couple simply because you can see the different abilities and interests that they have. My conversations with each of them are very different.

I'd like to hear more about your thoughts on how a product communicates. It seems that your interest in products as a form of language is at the core of what you and they do, but with different results.

As a society, we are unaware of how we grow up to read the physical world. It happens so early in our lives that we aren't able to articulate it particularly well in later life – even though it is fundamental to our understanding of how the man-made world functions. As children, we're brought up to sit in chairs that our parents provide and we're told not to walk into doors. Children instinctively know how a chair functions but will have no recollection of how that came to be. As we get older, we're happy to then take our weight and drop it into an object that we've never tried before, because it resembles the language of a chair – it must support our weight. What this says is that products are forms of belief. When you look at a product, you must fundamentally believe in it: it must be true. We gradually take for granted that the physicality of the world we live in must work. So, a designer finds themselves contributing to these truths in society, continually adding to a language, albeit a physical rather than a written one. It's powerful and also elemental: if you can start working out how to describe new things well, you can add to this language, and I believe that this is what Sam and Kim believe too. The conundrum is that if you try to create a new language – as most interesting designers do – how sensible is it really? Unless people have seen it before, or have experienced it, they have no reason to know whether it will work or not, or what it is, or whether to put their trust in it.

The enjoyment of any language is key too. For example, a set-top box for a TV contains a fairly complex navigation system, and is both difficult and unpleasant to use. Yet, the board game Monopoly is also a fairly complex system, but is used for fun. So why is it that one system could be so pleasurable and the other so frustrating? Perhaps it is because Monopoly is a memorable landscape of potential, a visual description of its system. I can read the changing state of the board and predict what might happen. The set-top box, however, lacks an enjoyable visual system and instead uses meaningless words like 'Audiodub'. So, product design has this continual dilemma, unlike other professions, that has to ensure that language and communication are familiar, but at the same time progressive, because there is invention to articulate. How can you visualize new things that communicate properly when the only reference point for people believing in them in the first place is if they've seen them before? This makes progress difficult, and it's a balance that Industrial Facility has spent a lot of time grappling with.

Remote pp. 137, 274

INFO
MUTE
SAVE
BANG & OLUFSEN

Jetlag pp. 137, 263

Timer Lamp pp. 138, 269

Cordless Phone pp. 138, 263

Tronco pp. 139, 275

Project Notes

Door Tool pp. *24–5*, 257

In the early years of the studio, there was less work and, in hindsight, this gave us time to think and apply ourselves to subjects that we were interested in, and in so doing, to create a kind of literacy through objects that set out our design philosophy. One of the first objects was the humble door handle.

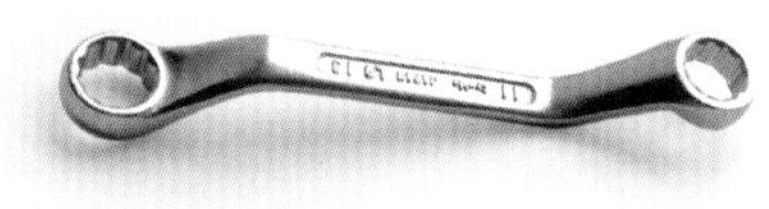

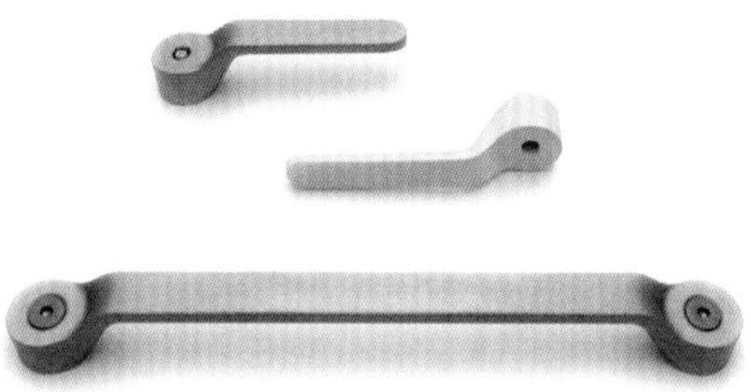

Sam had noticed that builders across the street were using temporary door handles supplied by the manufacturer during construction, until the 'good ones' could be fitted. They looked similar to a tool that slipped on to a square peg and, in our opinion, looked and worked better than the replacement ones. We thought that it would be nice to move this idea to greater prominence, and designed a handle that resembled a metal wrench, which, with its ergonomic contours, appears to grasp an exposed square bolt. Both the handle and peg visibly turn together. This project remains a good example of Industrial Facility's early works – an idea driven by initial curiosity (in this case, about how things work); borrowing rather than inventing; and using function as an aesthetic driver so that there is always a lightness of touch in the finished design.

Twinstopper pp. *26–7*, 268

Twinstopper is a compact refinement of a standard wedged doorstop. Commonly found doorstops are certainly a simple shape, but because they need to accommodate variable gaps between door and floor, their length often makes them protrude, causing them to get tripped over. Twinstopper is different in that it splits this wedge into a right angle, halving its overall length – with one side for small gaps, and the other for large gaps – that lets it stay close to the door. It is an acknowledgement not only of the door, but also of people walking by.

Droog originally asked for another design project as a follow-on from the rubber Flexlamp design [pp. *178–9*]. Knowing that Droog now had a good knowledge of rubber tooling, it seemed a sensible and easy route to propose a rubber doorstop. We have a lot of respect for Gijs Bakker and Renny Ramakers, because at that time Droog made projects that no one else would ever bother

with – odd projects that had some poetic usefulness around the home. Twinstopper remains useful (and in production) to this day.

Wireframe pp. *28–33*, 271

The Wireframe sofa is designed to be moved easily because its lightweight frame can be carried separately from its soft, contoured cushions. Steel wire – a widely available industrial material that makes structures both lightweight and strong – has a long history with Herman Miller, but had not previously been applied to such a large span. Using it allowed the frame to be half the weight of a typical sofa. This lightness makes for easier transportation: the three-seat version can be carried up a flight of stairs by one person and is easily manoeuvred through small doorways. Furthermore, the separation of upholstery from its frame ensures that maintenance is easy: if a cushion becomes worn or stained, only that part needs to be repaired. All of these qualities are rooted in a consideration for the modern condition and a recognition that few people these days wish to be burdened by heavy furniture.

During development, we had the opportunity to use a new suspension method called 'super-seat', where a stretched injection moulding creates localized comfort rather than traditional webbing, which can act like a trampoline. The visual result of this technology is also very beautiful, as solidity is replaced by utter lightness and transparency.

Wireframe was the first piece that we developed under the creative leadership of Ben Watson at Herman Miller. Ben had requested a very progressive position on all aspects of its production, where any compromise was to be avoided. This stance, combined with the company's long hiatus away from soft furnishings, meant a long slog of prototyping in several factories, both

in Italy and America, before the right quality could be guaranteed. The collection includes one-, two- and three-seat sofas, as well as ottoman stool versions.

City in a Bag pp.*34–5*, 259

In 2004 Muji Japan asked us to help their European division, for whom Christmas is an important time of year. (In Japan, however, Christmas is not generally celebrated, although present giving may occur.) We set about developing a series of children's toys with the goal of combining wry wit, natural materials and affordability. In the initial meeting, Kim spotted a small box of farm animals made from wood that Muji were sourcing and immediately explained to Kei Suzuki, then European Director, that it was not really Muji, because Muji is an urban company, present only in cities. Kim argued that Muji should be making cities, not farms, and asked whether this in itself was an interesting opportunity. Kim then made sure that each city contained iconic architecture and at least one very contemporary building so that the set would remain distinct from souvenir products. London has the Gherkin by Norman Foster, Paris has Ieoh Ming Pei's Louvre Pyramid, and Tokyo includes the Tokyo International Forum by Rafael Viñoly. Each city was put into a simple natural linen bag with the city name printed on it in Helvetica.

City in a Bag became an instant hit, thanks in part to our designer Ippei Matsumoto who painstakingly researched and crafted each piece. Only cities that have Muji stores are featured, so as the years went by, more locations were added, to include London, Paris, Tokyo, Kyoto, New York, Italy, Germany, China and Taiwan, alongside a whole collection of other toys and games.

Together pp. *42–5*, 274

Our first years of working with Herman Miller were as design advisors. Starting in 2008, we met up four times a year with Bruce Mau, Susan Lyons, Gary Miller and (later) Collin Burry, reviewing both development projects and company strategy with CEO Brian Walker, and a small executive team. It was a very unique situation for us, as usually we are on the other side of the table – but there we were, analyzing and commenting on other designers' projects as they were rolled out in front of us. Looking back, it took enormous courage for the directors to listen to our independent honesty, which may not have been what they wanted to hear. This is perhaps why Herman Miller has always been such a special company.

During one review, we heard of a need for a new method of manufacturing, distributing and assembling furniture, for companies that were searching for much lighter and more adaptable workplaces. When we got back to the hotel, Bruce's eyes were wide with excitement: together, we acknowledged the potential of design to address it, so there and then we committed to collaborate on a project to tackle this challenge, calling it the 'Together' programme. Then followed two years of research and experimentation, where we developed a table made from paper and steel, and which we assembled using suction. This was the big idea – that by ignoring the conventional method

of fastening legs to tabletops, we could redefine the tabletop to perform more functions and so free the structure to also do other jobs. The panel is the size of a door so that it can be shipped by FedEx – a self-imposed constraint. It can be written on, attached to walls and attached to legs to make tables or easels.

The legs were engineered with the minimum amount of steel in ratio to the strength required in all situations. The beauty is that with no tools, all parts can be repurposed quickly, responding to this new need for a lighter, more adaptable work environment. The problem was perhaps ignorance on our part: that leapfrogging of conventional assumptions and methods is not always what people expect, and the project was mutually shelved. Still, it remains one of our most satisfying collaborations, where design almost reached perfect equilibrium in matching a material solution to an emergent conceptual problem.

Picturemate pp. *46–7*, *52*, 262

Most designs start by thinking about functionality and also about communication: how should a product convey what it does? There is always an element of balance to this process, which demands total honesty. In the case of this photo printer for the home, it was first and foremost an acknowledgement of just how little it is actually used – once a week, once a month perhaps, and possibly for most people only on special occasions such as birthdays. If a company or designer remains ignorant of the sliding scale and its requisite honesty, it can result in constantly believing that your product is the centre of the world, creating a will to make it more and more captivating. This is why when you visit an electronics store, you are faced with rows and rows of visual noise and over-expression, which has little to do with the product's function and how often it is used. No manufacturer likes

to hear that their product may not be used or looked at every hour of the day, but Epson Design had enormous courage in accepting our suggestion of a more appropriate level of product communication.

At the time of developing the design, Kim was creating an approach that she called 'Product as Landscape', which pertained to products having visual 'anonymity' when not being used and becoming 'alive' when in use. This translated in to a portable printer as a kind of lunchbox with a handle, which allows it to be stored easily on a shelf or in a cupboard. Then, when you want to use it, the lid folds up, the screen pops out and the tray turns down. The handle became even more useful when Epson added a battery pack for printing in any location.

This was the first Epson printer created by an outside designer and manufactured. The Picturemate continues to have great commercial success, continuing to stay in production where normally a design would be produced for not much more than a year. It also created a new typology for electronics competitors, whereby we started to notice everything from portable DVD players to wireless speakers referencing the Picturemate's form.

Beauty as Unfinished Business pp.*48–51*, 275

Invited by curators Benjamin Loyauté and Elsa Francès, we both curated and designed an exhibition for the ninth Biennale Internationale Design Saint-Étienne in France. Entitled *Beauty as Unfinished Business* (2015), it comprised a carefully selected group of recently designed objects, which were arranged to show a version of beauty that reveals the relationship between a product and its context – perhaps the closest we could come to describing what we ourselves find beautiful. As Kim said, 'It's difficult to singularly locate just where beauty resides in an object: is it in its making, its use, its form or its cultural relevance? Or is it that beauty is found in the equilibrium of all of these things?'

In a large darkened space of 900 square metres (9,688 square feet), the exhibition pavilion is a brightly lit series of open rooms that suggests an imagined but non-specific architectural context for each object. Each 'room' gives generous space to each object so that little else is seen in the viewer's periphery; however, there are glimpses suggesting 'more'. Around the entire pavilion is a curved line of black ropes that don't quite touch the floor, giving a suggested circulation route, as well as an intimacy to each room. Ropes were spaced widely enough to still allow visitors to enter and exit anywhere – to find their own way of seeing.

Beauty in products is not only to do with the age-old unity of material and function. There is also a spatial dimension that is very necessary and rarely discussed: a product is nearly always experienced within a context that arouses something in us. This 'something' happens when we look beyond the thing itself, beyond its own space, to something much bigger than ourselves. The exhibition was visited by almost 200,000 people.

Circumference pp. *53*, 265

In our office, each designer has a small Shinwa ruler on their desk, which we buy from Tokyu Hands in Japan. The legibility of its custom typeface somehow makes measurement very fast and satisfying.

We proposed that Muji contact Shinwa in Japan to ask permission to borrow their typeface for a new watch that we were designing. Called the Circumference, it follows the concept that a wristwatch in its purest sense is an instrument that measures time – just as a ruler measures length.

With Shinwa's permission sealed, we used the typeface for the dial. To extend the idea further, we asked Seiko Instruments (who went on to produce the watch) to print the graphic numbers and increments onto the back-side of the glass crystal so that they float above the watch hands. Furthermore, we selected a green-tinted glass and made it proud by 1 mm (0.04 in) in order to give a greater sense of its materiality.

All of these small interjections work together to create a design that has an interesting sense of depth and simplicity, produced through the interplay of three elements: the numbers printed on the back-side of the glass; the shadows that therefore project upon the face; and the hands of the watch. Here again, design is not so much a creative act of form-generation, but rather a refinement that comes about when materializing an idea that is rooted in borrowed simplicity; in this example, the Shinwa ruler. In an early rendition, we developed a rubber watch strap, which had the same colour as an elastic band, but which was later dropped in favour of leather and nylon. Circumference has since inspired countless copies.

Fan pp. *54–6*, 262

Recognizing that a rotational cooling fan is often found in the middle of a room where all its sides can be seen, we decided to give as much attention to the back as to the front. It was technically challenging, as the rear fan motor and gear system had to be reconfigured so as to obtain a perfectly smooth cylinder. The rationale for this is not just for 'cleaning up' the product, but also to make a connection to the inherent roundness of the front and to the translation of motion in the rotation of the blades.

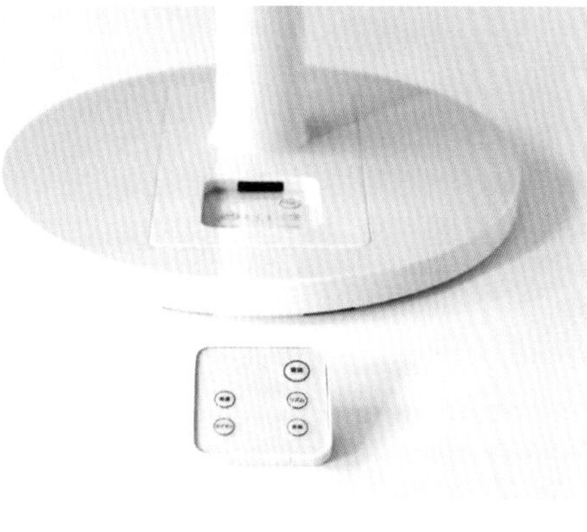

Sam visited the Royal Air Force Museum in London to take photographs of jet engines, which formed the basis of the concentric grid on the rear, 'to give the product power', in his words. The remote control was also a nice resolution: when not in use, it slots into the base like a puzzle piece.

Cycle Shirt pp. *57*, 270

The British fashion designer, Margaret Howell, approached Sam with the suggestion to design a shirt together, after her previous collaboration with the designer, Sir Kenneth Grange. More than just a shirt, it was a chance to exchange creative methods; and the surprise lay in just how similar our processes were, in our shared desire to pare away the inessential to locate the purely useful.

Being keen urban cyclists, and with the arrival of smartphones with smart wallets, the possibility of cycling with just a phone, a notebook and a pen has been made real, freeing the cyclist to enjoy the experience of the city while carrying very little. The Cycle Shirt promotes this freedom in its materiality – soft cotton that allows the body to breathe – and shape. We added a collar button, should it get draughty; a secret cardholder on the sleeve; and pockets on the back, similar to classic cycle shirts but without the sporty appearance. Margaret liked the idea of only one colour too – Sam's favourite powdered blue.

The pockets on the back give the shirt its main character; ridiculously simple, they are made by folding up the traditional bottom tail of a shirt, stitching the sides and middle, and then adding two of the same shirt buttons.

Pentagon pp. *58*, *60–1*, 270

The Pentagon pen collection consists of fountain and rollerball pens that represent what we call characterful simplicity: the idea that the actions of how an object is operated and used directly influences its visual character, without interference.

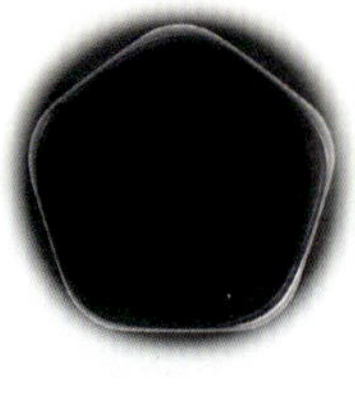

This can often result in an object appearing a little odd (some might even say ugly). In the case of the pens, the pentagonal barrel is held

like a bolt and the cap unscrewed clockwise. The cap can then be screwed over the barrel to transform the pen, hiding the barrel and creating a seamless design. It's all made from polished ebonite – a very hard rubber – that is shaped and formed by hand; and this hand-driven process enabled us to use an asymmetrical, five-sided form. Only a skilled craftsperson can produce it.

The process of the project was equally characterful. The company owner, Herman Chan, travelled from Singapore to our office in London, rang the doorbell and asked through the intercom if we had ever designed a pen. (In the office, we have a rule of thumb – if neither we nor the client have experience, then the project often fails. Failure also occurs if both of us are experienced. Generally, if one party is naive and the other is experienced, beauty is created. Mr Chan lives and breathes pens, so we knew we were in good hands.)

Passport Memo pp. *59–61*, 263

One of the things that Muji is best known for is its stationery range, which has become a staple ingredient and source of affection for many people around the world. The Passport Memo was created, developed and produced entirely through conversation, without a single drawing: a testament to the beauty of conversation as a way to describe an idea and test its strength. Sam returned from a trip to Japan where he saw everyone going through the immigration-check process carrying a passport, which is in itself a type of small notebook where information is referred to and collected on its pages.

A discussion in the studio concluded that a pocket notebook for Muji could share exactly the same colours, size, corner radius, number of pages and binding as a passport. The passport colour is an interesting element; we initially thought that it related to nationality or country of issue, but discovered that it is, in fact, relatively arbitrary (although countries seem to stick to four main colours – blue, green, red and black). The specification was sent to Muji, whose in-house graphic designers were able to find the correct papers and cover of a good quality, at a reasonable price. Gridded versions were also added later.

The project experienced success very quickly, perhaps because passports are themselves so familiar and generally well cared for. Sometimes, a designer's role is simply to be sensitive enough to reflect contemporary possessions back to us, but in a new use. Passport Memo continues to be produced (production having previously halted in 2015, which created such an uproar from customers that it was brought back): no doubt, it has become another staple item for many people, its convenience becoming part of the fabric of life.

Branca pp. *62–5*, 269

Mattiazzi is a small, Italian producer of wooden furniture, nestled in the Udine region, whose history lay firmly in anonymously making beautifully crafted pieces for well-respected Italian furniture brands. Going forward, they decided to presence their own brand by collaborating with a group of idealistic designers who would bring Mattiazzi the attention they deserve for their dedication to making exceptional wooden furniture. Up until our first project with them, our studio had not really tackled the subject of a chair, instead focusing more on mass-produced products, where good design was more rare. But, by 2010, when this project began, it was increasingly rare that the manufacturers we were working with had their place of production in the same location as the company's headquarters. The factory was becoming (for us) an at-arm's-length experience, with discussions held through agents and middlemen, whose job was more to do with cost reduction than with experimentation or appropriate quality. In this search for ever-greater 'efficiency', Industrial Facility has often had to argue for the value of quality and risk-taking.

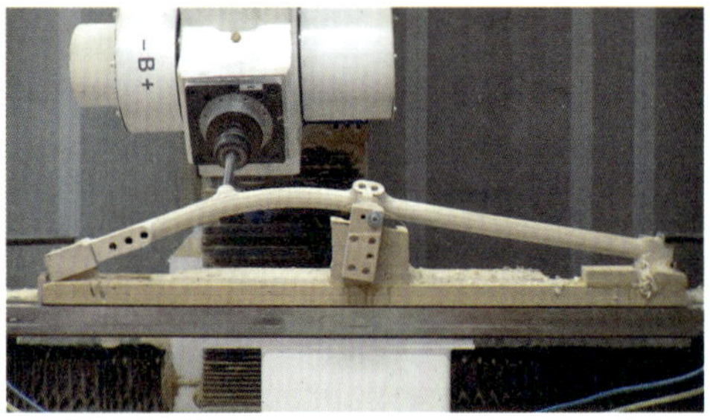

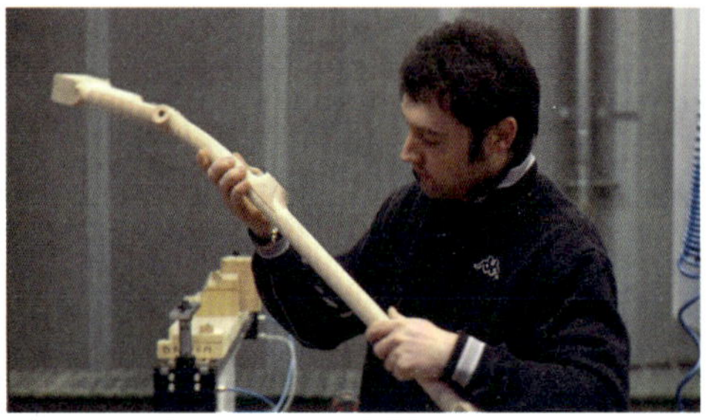

With Mattiazzi, however, things couldn't be more different. The Italian model – which sees the value of design as a cultural contribution, as much as for serving needs – is still alive in their part of the world. Our first visit to their factory showed them living and breathing production, taking up a position of adaptable craftsmanship, and understanding wood at a level that is only obtained through years of research and dedication. These first

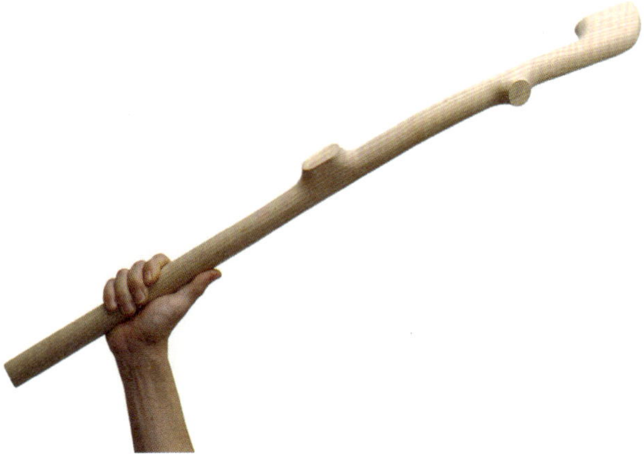

visits were set-up by Mattiazzi's then creative directors; Nitzan Cohen, himself an accomplished designer, and Florian Lambl, who still acts as art director. They showed us a factory able to turn logs into chairs entirely under one roof. As a family company, headed up since 1978 by the Mattiazzi brothers, Nevio and Fabiano, the only problem they faced was how to navigate their own potential.

The first project, Branca, combined our innocent naivety with their expertise – opposites that were (in hindsight) of profound importance to the development process. With Mattiazzi making no preconditions (other than that it be made of wood), we decided to harness the freedom of the CNC robot and the repetition that machines provide, and balance it with the skills of their craftspeople to form a synchronized relationship. In conversations between the two of us, the focus turned to nature, where complexity thrives with its own reason, and beauty results from the growth cycle. In particular, the branches of a tree provide the structural analogy for the project. Branca's back legs support the critical joints of its armrest, seat and back, and are made from a single piece of wood produced robotically. The joints are seen, but as a part of the seamless nature of the chair; and its simple outline belies the many steps in production.

As for the proportions and sculpting, this first took place at a scale of 1:10 and then 1:1, with models made entirely by hand prior to any computer modelling. The physical making of models is a key part of our process because if a chair is to contain the body and be grasped by hands, it should be designed by and with the hand first. An example of this is seen in the armrests: Sam grabbed a wooden broom from the workshop, measured it, and said, 'Let's make them 30 mm diameter.' A broom may seem a lowly tool, but the hand has held it for hundreds of years. For the colour range, Kim took photographs of forests in

the south of England: one of them showed a bright, moss-green trunk, and the photograph became the basis for the complete colour range.

We were so enamoured with the production process that we asked our friend, the film-maker Juriaan Booij, to record the complete journey from wood block to finished Branca chair. The short film, entitled *Branca: Something of How They Are Made* (2012), paved the way for a new appreciation of craftsmanship combined with new technology that embraced the making process as something to be celebrated. The Branca table was added along with the Branca stool in 2013. Branca has become an icon for Mattiazzi and is considered a fine example of how handcraft can work together with the newest robotics: neither of them is the same without the other.

Landscape Camera pp. *66–7*, 257

One of the earliest discussions in the studio centred around a shared awareness that at some point, many of the products that we use may not always require a person to activate every function of them. Products could no longer be thought of as having only a one-to-one relationship with us because the Internet would enable products to be networked so that they could talk with each other, without having to expressly talk with us.

We imagined that the context for products and their surroundings would grow in importance and, fifteen years later, we can now see this clearly happening in networked and voice-controlled products. In 2002, however, these thoughts were still in their infancy and we experimented a lot with breaking product conventions. We asked the question of what a camera would be if it could be controlled by the subject rather than by the person holding it? Our Landscape Camera did this by covering the front face of the camera with a two-way mirror film, behind which the camera mechanism lay. The person whose image is being captured would control the camera's position and, similarly, the camera could be left in the landscape with a self-timer. Its form appears reduced but still recognizable, with the shutter release also being used as an optional viewfinder.

Coffee Maker pp. *72–3*, 261

An opportunity arose at Muji to investigate an affordable, small-capacity, electric, filter coffee maker with an insulated pot. Japan has a liking for filtered, drip-brewed coffee as it emits a pleasant aroma in the home during the brewing process. As with most Muji projects, there was no formal brief – simply a conversation with a few slides. Our goal was to design the simplest of solutions for making good coffee and not to make a symbol of designer expression. We started by exploring to what extent this product should reflect its landscape of use, and it felt natural that, as most kitchen items are round –

plates, bowls, pans, cups – we could appropriate the roundness of form. The cylindrical design is also informed by where it can be used; it sits equally comfortably in a corner of the kitchen or centrally on a counter, something that most of these products fail to achieve.

The bent metal handle design was 'lifted' from a small, stainless steel *lattiera* (a milk jug) that we had found a few years before in Italy, kept because of its sheer rudimentary appearance. The design was first presented as a simple cardboard tube that purposefully ignored existing drip models, where the water container and brewing activity take place in separate volumes.

The design was accepted on the spot, a credit to Muji's ability to visualize the outcome and also to trust the logic behind its resolution. This was followed by realistic proposals for all of the mechanics, water filtration and, in particular, the rearrangement of a water reservoir so that it wraps around the filter holder, resulting in a perfectly circular footprint. Initially, the design of the component format was resisted; however, with some persuasion in the form of models and drawings, the factory agreed to try. A small, replaceable, charcoal filter that enhances the coffee taste was also added. An interesting note is that our first concept eliminated the heating plate altogether, which we viewed as superfluous for a pot that only holds two or three cups of coffee: however, this move was too reductive even for Muji. The design was commercially successful and remained in production for many years, with some small details updated in 2010.

IF4000 knife rack pp. *74–5*, 261

To accompany the IF4000 range of knives (shown opposite), a knife rack seemed an obvious follow-on project. The strikingly simple design consists of a single strip of laminated bamboo or oak that invisibly attaches to a wall, and contains six strong earth magnets. An unadorned block of wood attached

to a wall is an odd image, but it is made useful (and understandable) when a knife is attached. The knife handles remain easily graspable, floating a comfortable distance off the wall. The design sat perfectly with Taylor's Eye Witness, whose earlier works in the 1960s by the English designer and silver-smith, Robert Welch, also have this 'odd but useful' character, and still remain enduring. The knife rack became the company's most plagiarized design.

Chantry Modern pp. *76*, 257

It was in the 1920s that James Chantry invented a knife-sharpening concept of two rotating steel burrs (knurled rods) that are set at a precise angle to mimic the rotation of a butcher's steel running along a blade. Chantry spring-loaded these burrs to avoid excessive pressure being exerted on to the knife blade as it ran across them, which could lead to damage. To sharpen any knife, you simply run the blade between the steels four or five times – as if slicing bread – with tiny, serrated teeth applied to the blade.

Forty years later, the cutlery designer, Robert Welch, kept the internal mechanism but packaged it into a charming utilitarian product that you could screw onto a worktop. Another forty years passed and Taylor's Eye Witness, the company that had acquired the rights in the 1940s, commissioned us to modernize it as part of a competition run by the British Design Council. Called the Chantry Modern, our design moved the hand further away from the sharpening point, with a handle similar to the very knife that is being sharpened. The result is that the sharpener appears more like a cooking tool and less like a piece of machinery. After testing it during the project, we decided, like Welch, to not touch the internal mechanism, which had now become a replaceable part. The three designs still sell alongside each other – each with their own character.

IF4000 knives pp. *77*, 259

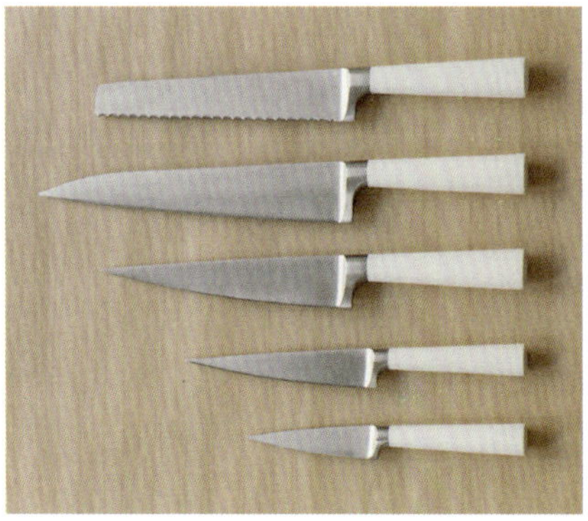

After the success of the Chantry Modern Sharpener, Taylor's Eye Witness then wanted the studio to look at kitchen knives. A chance conversation with a sushi chef in Japan inspired the direction that the project took: when asked why it is that you never see female sushi chefs, he replied that a man's body temperature is a little lower than a woman's. When handling the fish, it is always best to keep it as cool as possible for bacteria management.

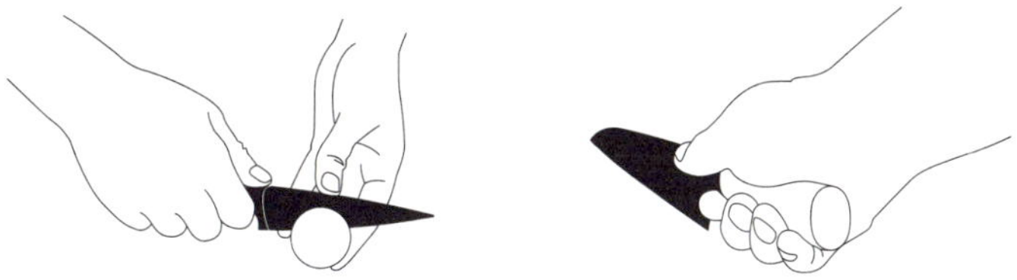

This fact sparked a search for a knife handle material that would encourage the hand to stay cool. A white polyester and melamine composite, similar to what dentist tools are made from, seemed a good material candidate. As well as this new material, the company adopted a new precision technique of forging and grinding, to create a smoother join between handle and blade.

A lot of contemporary knives are shaped to fit the hand with what we call 'pseudo-ergonomics', the claim being that this bias towards ergonomics somehow positions a form into the hand perfectly. The drawback with this approach is that it can sometimes limit the many different ways in which we might actually hold and manipulate a knife: while they may 'appear' ergonomic, it is more to do with aesthetical manipulation. The IF4000 handle has a deliberately simple, tapered oval, meaning that even with your eyes closed, you know which way the blade is facing.

Bottle pp. *78*, 274

Bottle watch is a unisex, analogue watch with a circle of raised nodules around the perimeter of its crystal, similar to those found on the bottom of a glass beverage bottle. It was simply one of those curiosities in life that we noticed that these nodules added up to sixty; and we made use of it as a method for timekeeping.

It's a graphic folly – no more than that – and by replicating the texture that these nodules make, it creates a striking iridescence when light hits the watch face at different angles. It was designed for Nava, an Italian accessories company, in colours typical of glass bottles – brown ale, green lager, clear vodka and blue water.

Wine Opener and Peg pp. *79*, 265, 267

It was Muji's request that one tool would be able to do two different jobs concerning a wine bottle; they asked for a tool that would both cut the bottle foil and also serve as the wine bottle opener.

The bottle opener handle appropriates the exact size and form of a wine cork itself and, in so doing, it becomes a convenient place to store the foil cutter. Around the same time, Muji also asked for a matching bottle stopper with a simple lever for sealing a bottle. Both designs reduce the ergonomic aesthetic commonly applied to these types of tools when they move to a plastic material.

Radice pp. *80–3*, 273

Radice stools are our second collaboration with the Italian furniture producer, Mattiazzi, and perhaps our most sculptural project with them. The three-legged stool, in both bar and chair heights, obtains its underlying beauty from its structure; by bringing together the front-half of a traditional four-legged stool and merging it with a single back leg – the root – that then spans across at the top, a modest backrest is created that is small yet reassuring.

The seat is deliberately pushed forward to give more prominence to this back leg and, when it is viewed from the side, it looks almost typographic. Like the Branca chair before it [pp. *62–5*], it capitalizes on the superb handcraft and care in manufacturing that Mattiazzi is known for. No screws or metal fittings are used in its production. Radice continues to be our most plagiarized design for Mattiazzi; copies appear in restaurants and offices but without representing the soul of its presence – the radii and angles that make you feel something. The wood stain colour options are based on the cycle of an autumn leaf changing colour.

Dalston pp. *84–7*, 275

Most of us are familiar with the outline of industrial pendant lamps typically found in warehouses. They are the metal ones hung in multiples that direct light downwards as well as softly illuminating their outer shades. While they once stood for practicality and durability, their relevance has diminished with the move to energy-efficient LEDs that can be programmed to have a controlled spread of light (the lamp's shade is somewhat arbitrary). Nevertheless, their iconic shape remains endearing and sometimes it's a designer's role to enliven old archetypal forms using new technology, with the challenge of integrating them into something that makes sense.

Our knowledge of LED technology from Busby [pp. *198–9*] led us to design a lamp module made from a colour-anodized aluminium casting, using it as a method to 'hang' either metal or glass shades, but programming the LED spread so that some of its light spread across the outside shade, too. The lamp module also acts as a diffuser, obscurer (from bright spots) and heat sink, to dissipate the heat away. When we shared the idea with Wästberg, the excitement in seeing a separated shade that hangs from an LED module was irresistible; they had found a supplier in Italy who could coat metal in any RAL colour at an economical price, allowing endless variations in colour combinations.

One thing that did surprise us was that the cast aluminium module could be anodized. We were always taught in school that anodizing cast aluminium was not possible because of material impurities. But Wästberg's factory in China found a way of using pure aircraft-grade aluminium alloy.

Alumi pp. *88–90*, 272

The French company, Tectona, known for their practical wooden garden benches, asked if we would be interested in the design of furniture for the outdoors. In hindsight, it was risky to ask a wood company to veer away from their most commonly used and sold material, but we had already worked a lot with wood, and we wanted to turn our attention to aluminium – a perfect material for the outdoors as it is lightweight, non-corroding and can be shaped easily. There was an idea floating about the studio to appropriate the typical handrail diameter found on public stairways, and apply it to the legs and armrests of a chair. Having a chair that appears chunky and solid, while remaining lightweight, appealed to us.

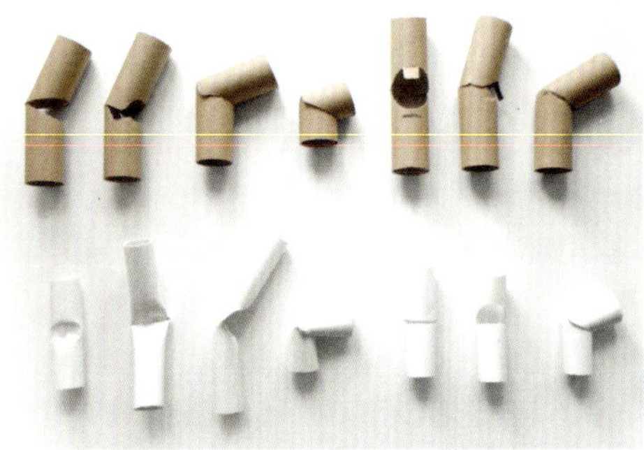

The thick arms and legs are made from a thin-wall aluminium extrusion. The large-diameter legs also contribute to the chair's overall strength, only needing two (rather than the more common three) crossbars, reducing the chair's price and maintaining good stackability. After countless studies of connections and profiles using cardboard and plastic tubes, we settled on a concave form that is cut in to the leg's profile, which allows it to stack six-high. To visually capitalize on this stacking, we selected colours that are tonally compatible, providing a very beautiful arrangement when they are stacked high or mixed together around a table. A flat-pack table design was also produced.

Paper Cup pp. *91*, 259

The first disposable paper cup was designed for use at water fountains in the early part of the twentieth century. It replaced a single-use metal cup, called a 'health cup'. Today, paper cups speak more about convenience than anything else, having no explicit relationship to their contents. Plastic cups are even more removed from any connection to what lies inside.

In response to a conceptual group project called 'Haptic', conceived by our friend, the graphic designer, Kenya Hara, for the Takeo paper company, we designed a disposable paper cup in two sizes, bringing something more than just convenience. It's saying that even something as nonchalant as a paper cup can be a little bit more sensual, revealing a hint of the cup's contents: a visible but undefined quality.

Milk appears different from orange juice, just as wine is distinct from beer. The translucency of Takeo's waterproof tracing paper allows this difference to show through, conveying a haptic quality that regular paper can't provide. For a product that has such a short life – sometimes seconds – it demonstrates that we shouldn't sacrifice the enjoyment that's typically found in objects that stay around for much longer.

Table, Bench, Chair pp. *92–5*, 266

It was while riding the Tokyo Yamanote line that we noticed that the seating on one of the trains appeared as long red benches, free of any armrests that would normally divide it into individual seats. To overcome the problem of seat assignment, the bench had one pink coloured zone – the size of a chair – so that the first person to enter the train would unconsciously gravitate towards this pink part and sit down. This first passenger had then effectively subdivided the rest of the bench for everyone else, so that no space was lost as the train filled up along its journey.

What if there was a bench that used the frame of a chair to create a sitting zone? We tried it out on a project for Established & Sons (originally, as a sofa). An oak plank acts as the bench, which can be of varying lengths, with a bent wood and concave soffit used to delineate the seat. The juxtaposition of a bench, a chair and a table is visually perplexing, but at the time, Sebastian Wrong, art director for Established & Sons, relished the controversy. He made

sure to have it crafted so well that it lifted it above any humorous association. Its presentation in Milan's La Pelota was striking: the dim lighting with bare wood made for a contemplative atmosphere.

Remote pp. *100–1*, 274

TV remotes are one of those products that people rarely seem to endear themselves to. Most manufacturers still view a remote control either as an opportunity to express style and material, or to spend no time at all on its form or on any consideration that it is a controller of a complex system. Remotes either become bigger to accommodate more buttons that we neither use nor understand, or they go pseudo-ergonomic in their form, like a piece of sports equipment. With Bang & Olufsen, enough investment *is* ploughed into this little object, in the knowledge that it is vitally important.

Our first duty was to sort out the button configurations so that they meant something, and this led to a remote control half the size of normal ones, but with buttons on both sides. On one side, you have only the buttons that you mostly use – power on, scrolling through preselected channels, adjusting volume and muting – all in a large size. On the reverse, buttons that are required to access the complete system are placed alongside an LCD screen. Using a gravity switch, each side is activated when it is facing up, while the other side is deactivated. The form of the remote projects one surface up at an angle of forty-five degrees, allowing it to be turned over or picked up very easily. While this form is visually odd, it is beautiful at the same time, informed by its functional criteria. This attitude to seeing function as a type of special relationship is what we felt oozed out of the early works of the designer, Jacob Jensen. After a year's work, the project was halted in favour of working in other directions.

Jetlag pp. *102*, 263

It's not often that there's a chance to develop a new electronic product from scratch. The Jetlag clock is driven not by its shape but by the nature of its interface. The idea came from the frustration of trying to set a digital alarm clock for travelling: the answer came during a trip to Tokyo, where we stayed at the very beautiful International House. The bedroom's clock was built into the headboard and hadn't been updated since the 1970s; and it involved two digital displays – one for the time and one for the alarm. If you woke up in the middle of the night, you could instantly see both displays without having to press any buttons. This idea made things simpler, even though it effectively adds more components to the product.

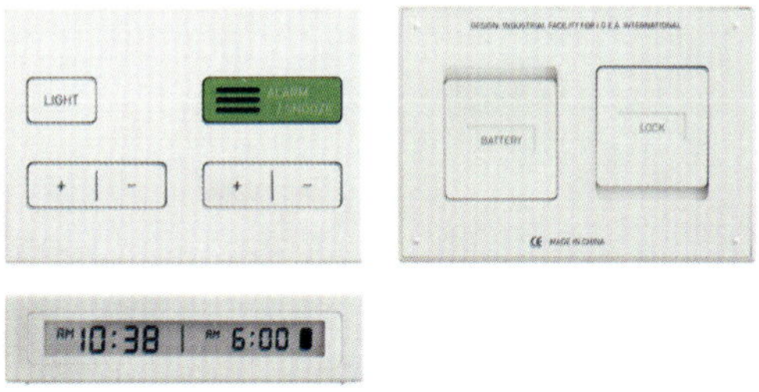

To transform this idea into a compact product required programmed PCB componentry – something that we had not done before, but the advantages appealed to the client and made the product utterly functional. The size is exactly half that of a standard passport. Not only are there two displays (the time display is slightly bigger than the alarm to create a hierarchy), but each display has its own buttons for adjustment on the corresponding top-side of the clock. The speaker doubles as a snooze button and the underside of the clock was influenced by Kim, who found an old Fisher Price toy that featured a simple sliding method for replacing a battery without the use of screws. The product met with success, particularly with older people who find layered interfaces that share the same button to be a source of frustration. In this case, more proved better than less.

Timer Lamp pp. *103–5*, 269

With the constant need for conserving energy on everyone's minds, Muji contemplated the usefulness of a night-time lamp that could slowly dim to an 'off' position as a child drifts into sleep. Using soft, unbreakable silicone, a uniform, warm light can be controlled by either an on/off switch or for a set time period of thirty or sixty minutes. The lamp gradually dims until turning off. The simple interface is discreetly placed underneath the lamp. Its form is influenced by small paper lanterns commonly found in Japan and is also a shape that feels natural in the palm of the hand when the buttons are pressed.

Cordless Phone pp. *106–9*, 263

After designing the Second Phone for Muji [pp.*200–2*], our attention turned to an idea for a cordless phone that could involve the component restrictions supplied by the same factory. Unlike the Second Phone, the cordless version is more recognizable as a form, partly because it makes sense to have it stand up and become an object in and of itself when away from its base. The idea of over-simplification carried over, however, so that clarity played out in material decisions: the surface that touches the face is a completely smooth, glossy and radiused surface. The necessary charging contacts are hidden inside a female void underneath, with the base instead containing a male peg that the phone sits on. There is only a small 2 mm (0.08 in) hole that gives in to technology (if we could have removed this detail, we would have). The reverse side of the face is where the interface is kept – its buttons and a small LCD screen. This format was opposite to that of cordless phones at the time: the advantage of it is that any oils from the face and mouth can be easily wiped

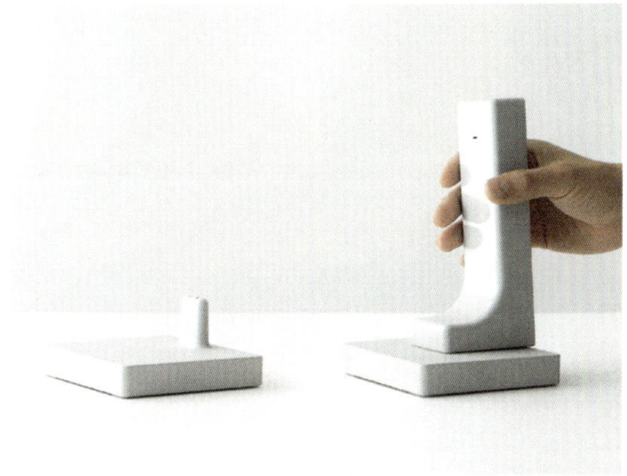

clean, and the glossy surface also creates much more of an 'object' rather than a technology impression. Investment was seen as too high at the time and the design was left as a concept.

Tronco pp. *108–11*, 275

The Tronco chair is unlike our previous Mattiazzi designs that display a sculptural softness achieved by robotically removing material. Instead, Tronco's character is formed from solid planks of wood joined together in what appears to be elementary construction.

What drove this approach was an initial conversation with Florian Lambl, Mattiazzi's creative director, who visited our office with a request for us to imagine a chair for a church. The questions that arise from such a condition proved the basis for the design: how could it work visually in multiples and not only for a single chair? Would it be able to form long, unbreakable rows? Could it stack ten-high and be easily set out?

Connected rows are achieved with a small moulded plastic connector that snaps onto the ends of the back legs. A custom trolley is available to help move the stacked chairs. Yet, these functional attributes are not visually present in Tronco itself – instead replaced by the appearance of 'texture', which lifts the atmosphere of interior spaces by creating a kind of wooden terrain. The frisson between solid panel construction and skilled craft makes for a humble presence, similar to some of Antoni Gaudí's church benches, which we had seen in Barcelona's Sagrada Família during the project. In 2017 an upholstered version was added, along with a table with chamfered corners that make them easier to abut in multiples.

Bell pp. *160–3*, 266

Even though the growing digital world has made the archetypical analogue alarm clock appear like a historic relic, people are still attracted to them for their loud sound and simple interface. Bell was designed to help this product take up a more desirable position; and to satisfy people who struggle with the layers of information required to set digital clocks.

The design enlarges the bell component and makes it the main part of the clock's body, helping to save components, simplifying the appearance, and creating a much louder ring. It features a snooze control, a moulded numeral display that avoids the need for printing and has LED illumination – when the alarm rings, the LED light also turns on, making it easier to see its display on dark winter mornings. The battery door is easily pulled off for replacement and also serves as a stand for greater stability when the alarm is ringing. It rings louder than a mobile phone, a desk clock or a watch. Its sound is also reflected in the choice of colours and their names: Fire Bell Red, Bicycle Bell Chrome and Doorbell Black.

Geneva HQ pp. *164–9*, 273

The Geneva HQ programme is a fully wireless, battery-powered security system that is easy to live with, easy to manage and ideal for self-installation. The range includes door sensors, smoke detectors, alarm sirens, a telephone that acts as a controller, motion sensors, glass breakage sensors, mobile phone charging zones and a security camera. All of these parts are suited to a variety of environments. The monitoring and notification system that layers over these parts can be made either through a security service provider, or directly between sensor and customer.

Sensible networked products are better when they are of service to their environment, and not just to us. We made sure that if a room could choose its own sensors, they would be ones that would blend into the background –

something that the majority of networked products fail to do. The security camera is a case in point. If you want to make a promise to a customer that it can be mounted anywhere and adjusted easily, you cannot then require it to be plugged in or expect everyone to drill holes at height. Our design appears as a simple round ball, like an eye, that runs on a battery and lasts about a year with normal use, because it turns itself on only when there is movement in the room. Its physical adjustment is by way of a magnet, so that all you see is a round ball with no screws. This magnetic adjustment feature went on to influence countless camera competitors.

Tubo pp. *163*, *174–6*, 273

The TOG company was set up with the ambition to combine industry with craftsmanship, using technology to thread it together. It was the brainchild of Brazilian industrial group Grendene, with Philippe Starck as its creative director. Our contribution is the Tubo collection of tables, shelves and a polyethylene (PET) chair.

The chair is compression moulded from a raw blanket material formed from shredded plastic bottles. Bizarrely, this waste PET is more expensive to utilize than materials that have not been recycled. We had already learned from past projects that the only way to achieve a reasonable price with a recycled material is to try to vertically integrate the process – in other words, to find a factory that is collecting, shredding, moulding and polymerizing (closing the cells using heat) all the material under one roof. Nicola Rapetti, TOG's dedicated design manager, was invaluable in achieving this. After a year of testing and analysis using a plywood tool, a steel tool was then developed that economically 'polymerized' and sealed the PET into a semi-rigid shell. The shell is then supported by a spring-steel exoskeleton base, which allows for comfort and movement. The result is a kind of 'woolly' seat that is warm and tactile, and can be customized with applied graphics using spray dye.

A worktable and shelving were also designed to complement the collection. TOG never did quite achieve its big ambitions, however. The company ceased trading due to a lack of distribution channels.

Lite Plug pp. *177*, 265

In the early years of the office, we wished to tackle products where design was heavily compromised, resulting in them being awkward to use or looking unfortunate. It was our choice to examine the world of emergency plug-in torches – a popular product in Japan due to the high incidence of earthquakes. The existing products are mostly illogical in how they are designed to be used, and are visually too brash for something that spends most of its life unlit and in a wall socket.

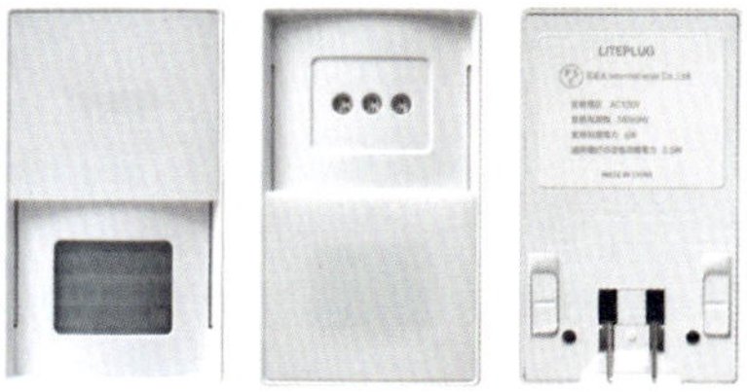

The Lite Plug plugs into a socket to become a night light, and the front corrugated lens slides down to reveal a motion sensor. As the room darkens, a single LED emits a soft glow through the lens. When someone walks past, three LEDs light up, making the room a little brighter. In an emergency, the Lite Plug senses power failure and all three LEDs turn on: it can then be pulled out and with the lens pushed up it becomes a bright torch. This was one of several projects for the small Japanese trading company, Takumi, with the aim of producing simple and straightforward solutions for people both young and old.

Flexlamp pp. *178–9*, 258

In 2003 the world of lighting had little to get excited about. The dominant compact fluorescent bulb, with all of its 'green' credentials, took time to warm up, had a sharp, cold light and appeared very unfortunate, far from the warmth of incandescence that we were all used to. We had made some experiments to try to improve the light characteristics and give it some quality of its own, drawing the observation that this fluorescent light could be considered to

be no more than an engineered filament, rather than the complete bulb that Edison had originally developed. We ended up suggesting a one-piece, silicone, soft moulded shade that shrouds the bulb, treating it as though it were a filament.

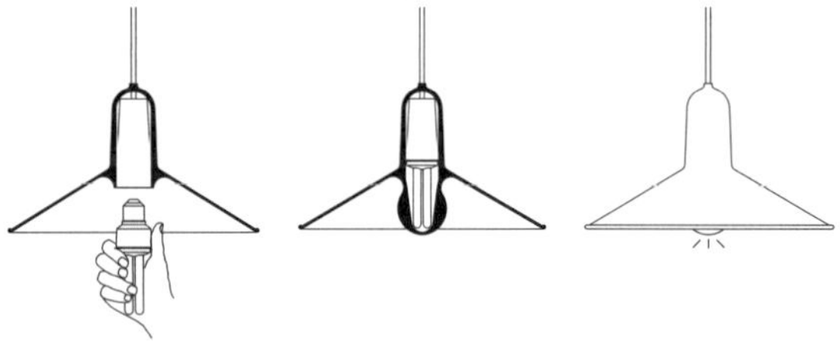

The project took some time to stick. First, Muji exhibited it in 2003 at the Salone del Mobile Milano (Milan Furniture Fair). Droog spotted it and asked if they could develop it for their own collection. Flexlamp then went on to spur countless derivative designs, until LEDs replaced its technology.

Semplice pp. *180–3*, 272

Each year, Sam would receive an email from Antonio Verderi, CEO of Oluce, one of the oldest Milanese lighting companies, starting with the words, 'Sam, I would like to make a project with you.' The only trouble was, Mr Verderi was sending the note to the wrong address – having meant to send it to the French designer, Sam Baron, instead. We'd politely reply saying that he'd got the wrong Sam. After about the third year of it happening, Antonio eventually said, 'Look, it may be the wrong Sam, but fate keeps repeating itself. Let's make a project together!'

Oluce has a special place in design. If ever there was only one table lamp left in the world, perhaps it should be Vico Magistretti's Atollo lamp. Its design permitted us to imagine eliminating the most characteristic of components for this type of table lighting, namely the lamp base and the stem, instead resting an inverted moulded glass cone directly onto a table surface. A further dematerialization effect is determined by the transparency of the glass itself. The lamp's electronics are held in a metal canopy within the top of the glass cone, allowing it to appear as though it were suspended in mid-air.

The deliberate reduction of these constituent elements brings with it a greater attention to details and finishes, in particular the rotary dimmer located at the top of the lamp, which is used at the same time for joining the metal and glass elements together. The mouse hole cut-out at the base allows the cable to exit. Rigorous, yet poetic perhaps.

Toilet Brush and Bin pp. *190–1*, 265

A toilet brush is not an easy thing to design well. To design is to consider first where this type of product spends its life, and then to examine how best it can be used. Designers are most often taught that it should be the other way around – to think of use first. However, toilet brushes are almost always pushed up against a wall and spend more of their life just sitting there waiting to be used. Most toilet brush designs ignore their surroundings, instead displaying personality, bright colours, fancy material combinations and so-called innovation – all meaningless concepts after it has left the shop. Simply said, we decided to make the form square with rounded corners so that it could find its home in a likely corner or close to a wall. We then made the base low and solid to avoid it toppling over, with a thin lip around the top edge that allows excess water that accumulates in the bottom to be poured away easily. For the bin, we matched the proportions of the brush container, and made sure that a waste bag could be held in place.

No one particularly wants to maintain these types of products once they are in use, so responsibility falls to the designer to get the details right for hygiene and maintenance, while still ensuring it sits quietly in the room. It has gone on to become one of Muji's most successful models and continues to be, probably because of its plain thinking and suitability for its surroundings.

Bath Radio pp. *192*, 265

Unlike a typical manufacturer, Muji has a freedom to apply itself to many territories of product, from food and clothing, stationery and furniture, to beauty and electronic products – always making sure that what is produced has a lasting relevance and an appropriateness to its setting. It was these last two territories – beauty and electronic products – that allowed Industrial Facility to pose the question as to whether a bath radio could be a simple extension

of their successful refillable bath wash range, which used a combination of high-quality bottles with a replenishing sachet system.

The radio 'borrowed' the form of the refillable shampoo bottle. The speaker also relates to the size of the shampoo sticker. Radio controls were carefully considered in terms of use – the on/off and volume control are at the top, and the channel selection is underneath: this acknowledges that the radio channel is rarely changed in a bathroom. The sound quality is enough for the purposes of a shower or bath and, with water-resistant sealing, it is not affected by moisture, like regular radios.

The connection between a radio and a shampoo bottle might seem absurd at first, but if both are viewed as things that we use during bath time, it is perhaps quite relevant. It was certainly a fine line: one that Industrial Facility have continued to walk ever since their Second Phone for Muji [pp. *200–2*], some ten years earlier.

OH! Toothbrush pp. *193–5*, 275

OH! makes the electric toothbrush a little more everyday in appearance, easier to travel with and allows the ordering of replacement brush heads from the product itself. It's the first electric toothbrush to combine charging and ordering technology with analogue practicality, and it deliberately removes the LCD display, with no data gathering and no cloud sharing.

We combined the brush with an app, enabling it to pair with any smartphone using Bluetooth. When new brush heads are required, the handle cover can be removed and an order button pressed, 'pushing' a smart reminder to your phone. For charging, it adopts the same reliable induction method, but for travel there is a built-in USB plug, so the charger dock can be left at home.

Braun had invited us into their headquarters in Kronberg im Taunus, probably expecting us to suggest something about craft – to fit into the idea of handmade beauty. But we had other ideas; for what makes a thing of beauty is a rounded sense of logic and simplicity. That means confronting the Internet of Things – both software and hardware – with a keen eye on removing all superfluous novelty. It was so clear to us that all products will at some point become connected that it led to a new department in Industrial Facility – Future Facility.

Little Disk pp. *196–7*, 263

Unlike other hard drives at the time, we wanted to create a collection that required no extra power leads or transformers, no buttons or display screens, no feet that would express orientation, no vents and no large flashing

LEDs. You simply plug it directly into your computer with its built-in cord. The premise for pushing for this was very simple – that a hard drive is not really a product at all but rather a piece of media, just like a CD or DVD (no one calls a DVD a product: it's something that goes 'into' a product). Each drive is of a different size depending upon its capacity and, using the concept of appropriation, the physical sizes picked up on dimensions that are familiar to us – a matchbox, a cigarette packet, a CD case, etc. We also redesigned their company logo and applied 'Super Graphics' that simply stated each product's specification – a device that Kim lifted from architecture to explain a building's purpose. For the larger, less-mobile drives it was felt that these should have an orientation, so we appropriated the method used on the bottom of toy cars that have a raised motif to illustrate the mechanics inside.

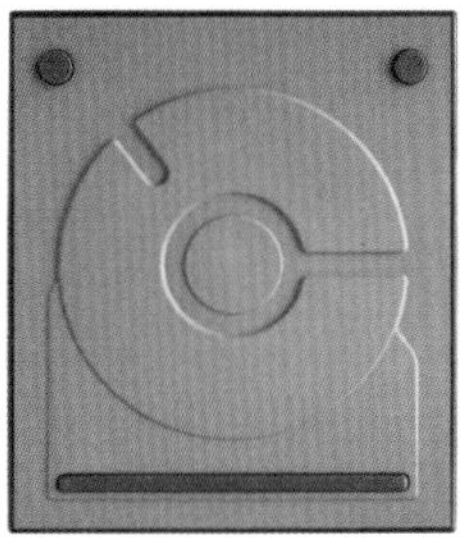

The concept presentation in Paris was memorable. In a room of about ten people, Sam placed beautiful models that replicated exactly the size and weight of each drive, and explained to them that a hard drive is not a product but a piece of media – so it should be vacant of all detail. At the end of the presentation, CEO Philippe Spruch went around the room asking for opinions. One after another, each person said that it was too simple, didn't have enough detail or expression, or would be expensive to mould. But Mr Spruch stood up and said to everyone, 'Great, that's exactly what I was hoping for!'

Development commenced the next day. The collection went on to become commercially successful, with photographers in particular admiring its no-nonsense, solid-state feel.

Busby pp. *198–9*, 274

For a long time, lamps have been an electrical experience – the light being evidence of electrical current passed through a filament. But modern lighting is no longer a subject of electricity, but of electronics.

Printed circuit boards, microchips, diodes and interfaces are now the staple ingredients, giving us greater control, conserving energy and providing longer lifespans. This transformation is a chance to establish new possibilities for the meeting point between light and electronics: wherever light is supplied, power is often also required to charge our burgeoning devices. Busby is an electronic lamp that provides intelligent, universal power from three USB outlets, detecting charging requirements and managing them at their fastest rate.

The project first started in a configuration similar to an extension cord, with an LED lamp plugged into one of its USB sockets. It was this prototype that captivated Wästberg and propelled them into the world of electronics. Gradually, over a two-year period in Sweden, this little lamp became ever more intelligent, growing into two versions – directable and ambient – and it remains one of our most enjoyable collaborations. We developed a striking animation with our friend, Johnny Kelly, and an equally beautiful catalogue with Ana Verala.

Second Phone pp. *200–2*, 257

The Second Phone has undoubtedly been one of our most important projects – not in terms of sales, inventiveness or, for that matter, functional ease – but in its radical simplicity that was made accessible to the customer, something that Muji is prepared to do. It is a telephone that intentionally has limited features and allows only basic communication, spurred on by landline phones becoming secondary devices to the more primary mobile phones. It relies only on one wire, needs neither electricity nor base station to perform and is activated when picked up from any table surface; the microphone is

embedded into a large 'engage' button that springs up when the phone is lifted. These are not so much 'selling' qualities, but when put together they create an object where comparisons to other telephones become nonsensical.

The thinking behind the project was part of the studio's early work in the field of simplified electronics, where a product relies upon its surroundings to fill in the gaps of functionality that the object may not have. If we know that it's a tabletop product, then why not get rid of the base and use the tabletop? (We tried this idea also in the Muji Coffee Maker of 2006 [pp. *72–3*], but it was a step too far.) This thinking leads to reducing the telephone to such a simplified form that all you can do is make and receive calls, and nothing more. The phone's secondary status is then made explicit.

When we first presented it to Muji's director, Masaaki Kanai, he had come to visit us in London. Instead of showing him the prototype in the studio, we took him to our home, showing how it could sit on surfaces and mount onto walls. The context of simplicity became very apparent; and he said with his usual directness, 'Let's make it!'

Projector pp. *203–5*, 263

A productive relationship with Epson Japan led to a project to design an LCD projector with an integrated DVD player. At the time, these projectors, unlike their cinematic forebears, had not managed to form a clear typology; instead becoming lost boxes in a world of changing lifestyles. Our approach first acknowledged the need for the transformation of a product, which was originally used exclusively as business equipment, into a useful and relevant tool for the home. Having an integrated DVD player and speaker meant that the projector had less need to rely on other equipment to function and could be a standalone product, rather than a peripheral one. All it needs is a white wall and an electrical outlet. The design's cine-projector reference is obvious – and forced the LCD componentry to be configured upright rather than flat. The remote control is in the shape of a DVD disc so that it can be stored flat on the product; the plug and cable fit neatly into the side (we repeated this detail in a hairdryer, a few years later); and the handle can be popped out as and when needed.

The result is a product that feels as though it was meant to play movies, instilling some pride and enjoyment through its use. Seeing the spinning DVD also contributes to the 'atmosphere' of playing a film. The design remained a concept, but has been influential nonetheless.

Store pp. *210–11*, 276

Echizen Urushi lacquerware is a tradition said to have started roughly 1,500 years ago in the foothills of Sabae, a prefecture of Fukui in Japan. The technique has evolved continuously, with advancements in material and technique to reflect changing times. Nowadays, the durability of what is called 'plastic lacquerware' – moulded plastic, coated in black and red chemical paints – has met the growing demands of the hospitality and airline industries. Almost eighty per cent of this plastic lacquerware is produced in Sabae, and Sekisaka Shikki is one such factory, having been founded in 1701. Japan Creative, along with Sekisaka Shikki's current leader, Tatsuhiro Sekisaka, approached us to see how a western designer might invigorate the company through what might best be described as a 'collision of cultures'.

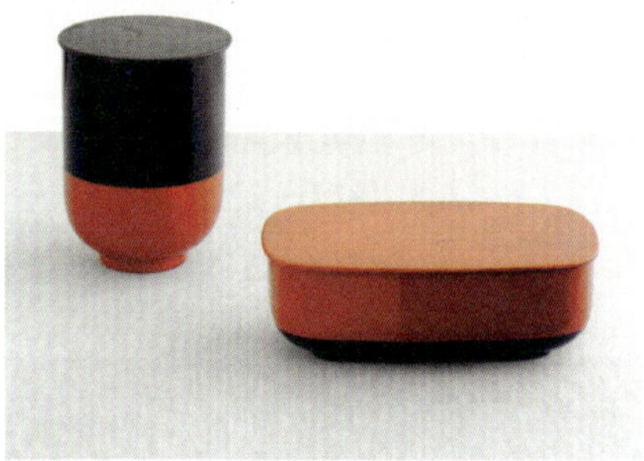

Our suggestion with Store is an appropriation of traditional Japanese rice and miso bowls that are already sized for the hand. The design lengthens the walls of the bowl 'up' to form containers with lids in different heights. They can be used for holding flour, coffee or anything that needs a dark place. The containers are made from injection-moulded plastic and coated by hand, either using spray for everyday use, or the traditional Urushi technique for special items. To further play with this appropriation, the Urushi lacquer version is two-toned, revealing a line of division where the rice bowl once was.

Before we started the project, we reread the book, *The Unknown Craftsman* (2013), by Sōetsu Yanagi. It is very telling that craft in Japan is not something that is viewed as elitist or special, but rather is a key part of the fabric of life. Modern Japan continues to find ingenious ways to save the knowledge of crafts from the constant onslaught of 'fast and easy'. For Sekisaka Shikki, the Japanese rice bowl form is a little sacred and, in truth, the idea of lengthening it to form containers is only likely to be suggested by a western designer.

Run pp. *212–15*, 274

Run is a collection of furniture and accessories inspired by the congenial atmosphere around public furniture, like canteen tables, park benches and library shelves. These typologies have the ability to feel natural for one to use and for many to share. A bench, much like a table, is a surface on legs; yet, somehow, its separated planks give a feeling of endless continuity that makes them less formal and easier to share. The materiality of Run consists of recycled aluminium that is extruded, and Amish-crafted ash wood. They are pure utility – resilient in material and timeless in form.

For the tables and shelves, specially designed accessories also made from ash help to delineate spatial and functional divisions along the surfaces;

the gap between the planks is taken full advantage of, making the bookends appear as though they are floating. For the table legs, we developed a new type of leveller that is simply rotated rather than unscrewed; and the whole collection can be used indoors or out, by way of clever engineering that disperses rainwater through the aluminium castings.

Gregg Buchbinder of Emeco was instrumental in allowing us to exercise what we call 'richness in reduction'. It was the first time that they were to extend their portfolio beyond chairs and, while project development will always have its wobbles along the way, Gregg surprised us with his combination of objectiveness while still allowing surprises to be fully taken advantage of – even up until the very end. When the furniture collection was completed, we photographed it in Milan with Miro Zagnoli. But Miro had put the table legs onto the seat by mistake. We all stared at it thinking, 'That looks like a fine compact table' – and it was duly added to the programme.

Two Timer pp. *216–17*, 265

With a steel frame and quartz movements, the Two Timer clock displays two time zones sharing the same dial, visually addressing the needs of the frequent traveller, the international banker, or simply someone with family or friends living abroad. Many cities, London included, continue to experience a lot of change as people become more transient. These days in a city, it almost feels like everyone has come from or is going to somewhere else. The Two Timer clock acknowledges this contemporary condition in a design that became very iconic for the British brand, Established & Sons. Its genesis was like so many projects: of their moment. Mark Holmes, co-founder of the company, visited the studio about a potential project and Sam took him for a beer at the nearby Jerusalem Tavern. With so many languages being overheard in the background, Sam quickly proposed an idea for a clock that could show two time zones but share the same dial; and Mark loved it. Surprisingly for a clock, a lot of manufacturing resources were thrown at its development, with an aluminium pressed housing and an extensive search for a printing method for a well-executed, large glass dial. At the time, Alasdhair Willis, another co-founder, suggested that the proposed size was too small and should be 'beyond what one would expect' for a wall clock. A 60 cm (23.6 in) size was added, in addition to the smaller 30 cm (11.8 in) version.

Locale pp. *218–21*, *226–7*, *230–1*, 271

Up until Locale, the largest projects we had tackled had been exhibition designs, and the longest in development had been no more than a couple of years. Herman Miller saw how the office was maturing and commissioned us to design a complete furniture system that improved the comfort, health and performance of working. It required extensive research and design that amounted to three years of development, some 100 components and

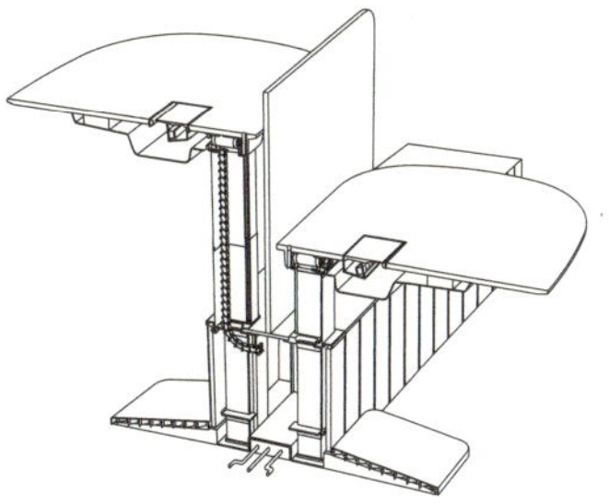

a team of thirty engineers and developers. Locale was also the first product we were to design that would become a complete environment.

The ambition we set ourselves was simple: our premise was that the joy of working in an open-plan office is that you are with other people and that working together collaboratively makes the work that you do more enjoyable. We know this from experience of our own office: the whole point about coming in is to 'create' not in isolation but with others. The sharing of ideas needs to be utterly opportunistic and spontaneous: the creative act cannot be planned to take place at a certain time. And so, when we hear of workers holding a creative brainstorm in a meeting room, it is in some ways nonsensical, for few of us can determine when our most insightful thoughts will come.

So our attention turned to the working desk, where working with others is the most natural – transforming it into a meeting table. Open, rounded and cantilevered, Locale abandons table legs (which may prevent people from changing positions), and instead allows two or three people to gather round easily. Desks can also go up and down to allow for standing, so extra chairs become unnecessary. The structure, mechanisms, motors, cables and storage are all encapsulated into what we call 'the workbase', helping to keep visual noise to a minimum. The complete system is modular, yet not visibly so; instead creating an architectonic language from which rounded surfaces are attached. Alongside this is a collection of mobile, adjustable tables, easels and screens that can be rolled around easily.

But Locale is not simply furniture that is systemic: it is also a proposal for a new spatial condition of how offices can be laid out; giving the world's top companies (who spend half of their time collaborating) a more relevant environment. It enables the creation of dynamic clusters of people that can form their own neighbourhoods in close proximity to one another.

When Locale was first launched, it was seen as radical, because the typology that it referenced was one for sharing – round meeting tables – rather than the individual, who is used to rectangular surfaces. But the reality is that the collegial atmosphere that Locale allows has led to a much more productive and joyful environment.

Formwork pp. *228–9*, 273

Formwork storage containers are designed to help bring some order to our desks. Each container is made from ABS plastic injected at a high pressure, to achieve a superior gloss finish, and mounted to a non-slip silicone base. They range from cups and trays to boxes and media stands, and can be stacked. The project is an affirmation that the things that surround us now are a mixture of the analogue and the digital; and that the true nature of what serves us on a daily basis is far more varied than what we might previously have imagined. Plastic spoons sit next to tissues; pens sit next to wallets; sticky tape sits next to USB sticks. A pencil cup may contain pens and scissors, but also inevitably SD cards, rubber bands, brushes and keys. You soon realize that the desk is an amalgamation of the office, the bathroom, the kitchen and the workshop.

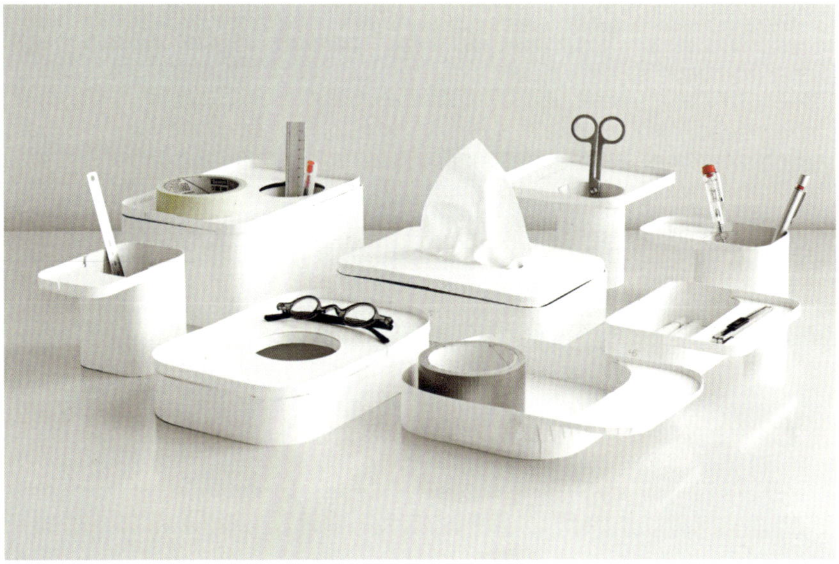

Over a period of time, we started to catalogue actual items found on typical desks, similar to how an anthropologist might examine life; and then started to group them. This avoided any attempt to beautify the situation, and instead arrives at forms and sizes of container that pertain to what they may find themselves holding. For instance, a Formwork cup shares the same diameter as the inside of a roll of masking tape.

It was very brave of Herman Miller to entertain such a project, where investment was high and the result was so simple. Imagine for a moment the presentation to management. 'We would like to design a series of simple plastic boxes for helping people to sort things out a bit better. We think that you have the legitimacy to do this.' It could so easily have been seen as frivolous and low priority. Instead, they agreed that even the largest of companies needs to also address the small things in life.

Shredder pp. *232–3*, 265

Around 2007, after working extensively on domestic products, something started to become awry with price. Imported kettles, drills and fans – the things that make daily life easier – had become so low-cost as to warrant them commodity items. An electric toaster was now cheaper than a good loaf of bread, which ultimately meant that meaningful design struggled to play a part in this territory, because it was seen as added cost. At the same time, furniture companies had started to approach us with projects that took us to a greater area of domesticity, giving us the opportunity to make some provocations for products from a furniture standpoint.

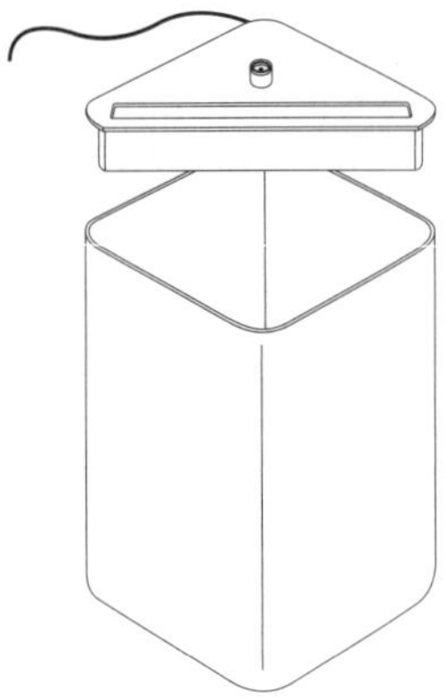

Shredders, in particular, were (and still are) cheap and unfortunate, and on a scale of furniture. As an exercise, we leveraged Gino Colombini's square waste bin, designed for Kartell in the 1960s, as the holder for the shredder component. Being square with radii allows it to hold the shredder diagonally so that it fits securely. It also means that it can be used as a regular paper waste bin at the same time. Only the cable and switch give away its function; otherwise, its geometric blackness suppresses any association with a plug-in product.

Bell pp. 140, 266

Tubo table pp. 141, 273

Geneva HQ pp. 140, 273

Magnus Wästberg with Anniina Koivu

I'm curious to meet Magnus Wästberg, CEO of the lighting company that bears his name. Located in Helsingborg, one of Sweden's southernmost cities, the eponymously named lighting company Wästberg is a business that is clearly resourceful, ambitious and utterly uncompromising. Founded less than a decade ago, it has been able to leapfrog much of the competition, with a fresh eye informed by a long family history in the lighting industry. At a rendezvous in Cologne, I spot a man in his early forties with curly hair that bounces in front of his spectacles, elegantly dressed in a three-piece suit. I want to understand his decision to work with Industrial Facility, where Sam and Kim have pulled Wästberg into radically new directions; and where, along the way, their relationship has changed, from designer–producer, to a friendship between two families.

Anniina Koivu: Magnus, you launched Wästberg in 2008, with an interesting group of designers, each with their own voice. How did you hear about Industrial Facility? And what were you hoping to find there? After all, they were not exactly known for lighting design.

Magnus Wästberg: I had them on my radar for some time. Then, I saw the Locale office system [pp. *218–21*] that they had developed for Herman Miller and it tipped the balance. It was the most brilliant office solution that I had seen for a long time. After that, Industrial Facility was an obvious choice for me to work with.

That project was a complete reinterpretation of work – the idea that the meeting of minds happens at the very place where you work and not just in meeting rooms. And while radical, it was a realized project.

Exactly. It wasn't just a statement or a question of restyling. The Locale system was a thought-through idea of how people actually work – or would want to work. To me, it was the best evolution of the workspace that I have seen – so refreshing and so relevant at the same time. That's exactly what I told Sam when I called him for the first time. I wanted to get to know the people behind Industrial Facility. I had a hunch that they saw design as a tool to articulate new potential.

It sounds like they hit a nerve with you, as far as how you perceive our built environment.

Yes, their projects speak of freedom, of liberation from the constraints of dogmatic design. That's also what I have been preaching ever since we started Wästberg. Initially, I had an idea to work with them on an office lamp, based on a system like Locale, which could adapt according to the context of use. We are still thinking about it – but, so far, it hasn't happened.

Instead, they proposed the multifunctional Busby lamp to you.

I would call Busby [pp. *198–9*] an evolution of what we imagine a lamp to be. It is not a frivolous gadget, but rather an affirmation of how we live now. Wherever light is supplied, power is also usually required – to power multiple devices from our phones to our laptops. We're tapping into that by providing power and light in a properly articulated design.

So was Busby love at first sight?

Definitely. I enjoy love at first sight! But then it had to be turned into a product. It took us a lot of time and investment. It was something new – in some respects, too new. Even if a lamp today is an electronic product – and no longer just an electrical one – and even if we have developed the know-how, and supply chains within electronic components and solutions, we were still looking for components and solutions that were not available on the market. So we ended up developing them ourselves.

Are you the referee in such a process?

To some extent, yes, but at the end of the day, what happens between the company and the designer is a long conversation. That's what I love about my job, this process of taking ideas and solutions, and turning them into physical objects. When I was a kid, we would drive from Sweden through Europe to Spain for our holidays every summer. On the way, my father would meet designers and visit factories, and I would join him. In the first year, we'd see the designers' drawings and I'd listen in on the conversations. The next year, we'd see a prototype. And eventually, there would be a final product. It's an enjoyable process to witness the creative act developing from an idea on paper to a finished product.

I would now like to talk about risk – something that you seem to willingly accept with your projects.

Please, talk to me about risk!

In some ways, your methods seem closer to the early Industrialist Italian model, like Fiat or Olivetti, whereby intuition and gut feeling are as important as research; and in this respect, a lot of investment ends up being put into unproven territory. Enzo Mari once affirmed this model by saying, 'If you want to eliminate risk from design, then your only real option is to copy.'

I love that quote. I absolutely agree.

So, is the work that you produce what you like personally?

Absolutely. But please bear in mind that that is very different from a question of taste. A product must be viewed holistically in relation to its long-term context. Being an entrepreneur, I am sometimes perceived as being a little stupid or irrational. You do things that you wouldn't normally do if you think beyond short-term numbers and surveys. Yes, I probably appear stupid to some people – the weird kid on the block – but I am confident that a long-term view pays off. So, risk is something that you have to be comfortable with if you want to be original. Maybe being stupid and original are one and the same thing.

Let's take a look at the Dalston lamp, from 2016. It is a clever pendant lamp. It feels so obvious, like something that's always existed.

I told Sam that I didn't think that pendant lamps were conceptually developed in a way that exploited the potential provided by LED technology. Our discussions ended up at the point of what could be described as a 'universal pendant lamp', reminiscent of lamps commonly found in workshops or warehouses. The lamp's genius lies in the fact that the quality of the light

remains the same, regardless of the shade's shape, material or colour. The shade can be painted to match any interior, which makes it incredibly simple and diverse at the same time.

It is an example of hyper-customization?

Yes, but to me, it is first and foremost a continuation of an ongoing tradition. It is a fixed point – one that you recognize, that you feel comfortable with. Dalston brings comfort in its appearance and in the light that it gives.

Sam and Kim define their work as 'a duty to ensure a project has not just commercial credibility, but also usefulness, appropriateness and longevity.' While that may be the studio's ethos, how would you describe them as people? Have you ever seen them angry?

No, but I have seen Sam worried. He thinks a lot – beyond what any client might imagine. Maybe, just like me, you need to get to know him. They both have strong characters. Kim is more outgoing and vivacious. She comes from architecture. She is the contextual person. Sam is more into the details. When you combine these characters, you get a fantastic result that is well-rounded and thoughtful. What impresses me most is how they work and rework an idea, over and over and over again. It is very seldom that designers rethink what they do to such an extent. People usually deliver a project and finish. But Sam and Kim always question and come back asking, 'Maybe we did this wrong? Maybe we should rethink this?' It never stops. They understand that what is technically possible is not always humanly preferable. They are thinkers. It is not about doing 'more' stuff, but about finding the relevant solutions.

It sounds like you've really got to know them.

Yes, it's a little like peeling off layers from a lettuce: with each layer peeled, the relationship becomes deeper. At first, it was through a series of dinners together. Then, Sam, Kim and their children came to our house in Sweden. Now, I would say that the client–designer relationship has morphed into a friendship.

Is it something special when designers come to your home? Or is that part of the deal when working with you and the family business?

It is a bit special. But this kind of relationship is something that I've come to treasure. It reminds me of an exhibition some twenty years ago that I saw at the Louisiana Museum of Modern Art in Denmark. It was about the Danish designer, Poul Kjærholm, and it was split into different rooms dedicated to his furniture. There was this one room that was all about his relationship with different manufacturers, with lots of stories about his close friendships to the people that he worked with. If you work with people that you like and respect, it is inevitable that some of these working relationships will turn into friendships. So, I try to work with people that I love. I work with my wife, my father and my father-in-law, and it's pleasant when you work, live and breathe with the people that you would also like to have Sunday lunch with. Sam and Kim: I regard them as extended family.

Tubo chair pp. 141, 273

Lite Plug pp. 142, 265

Flexlamp pp. 142, 258

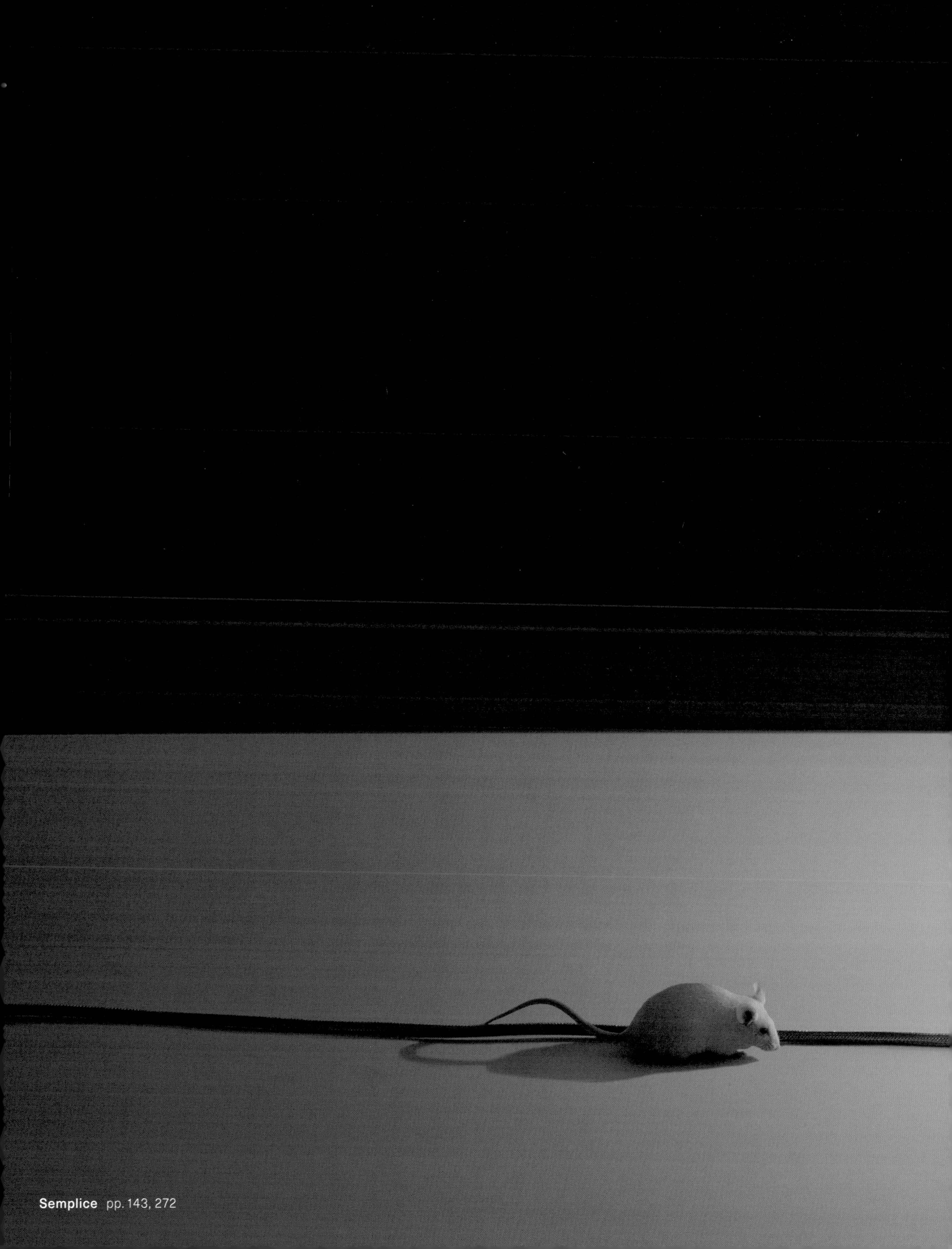

Semplice pp. 143, 272

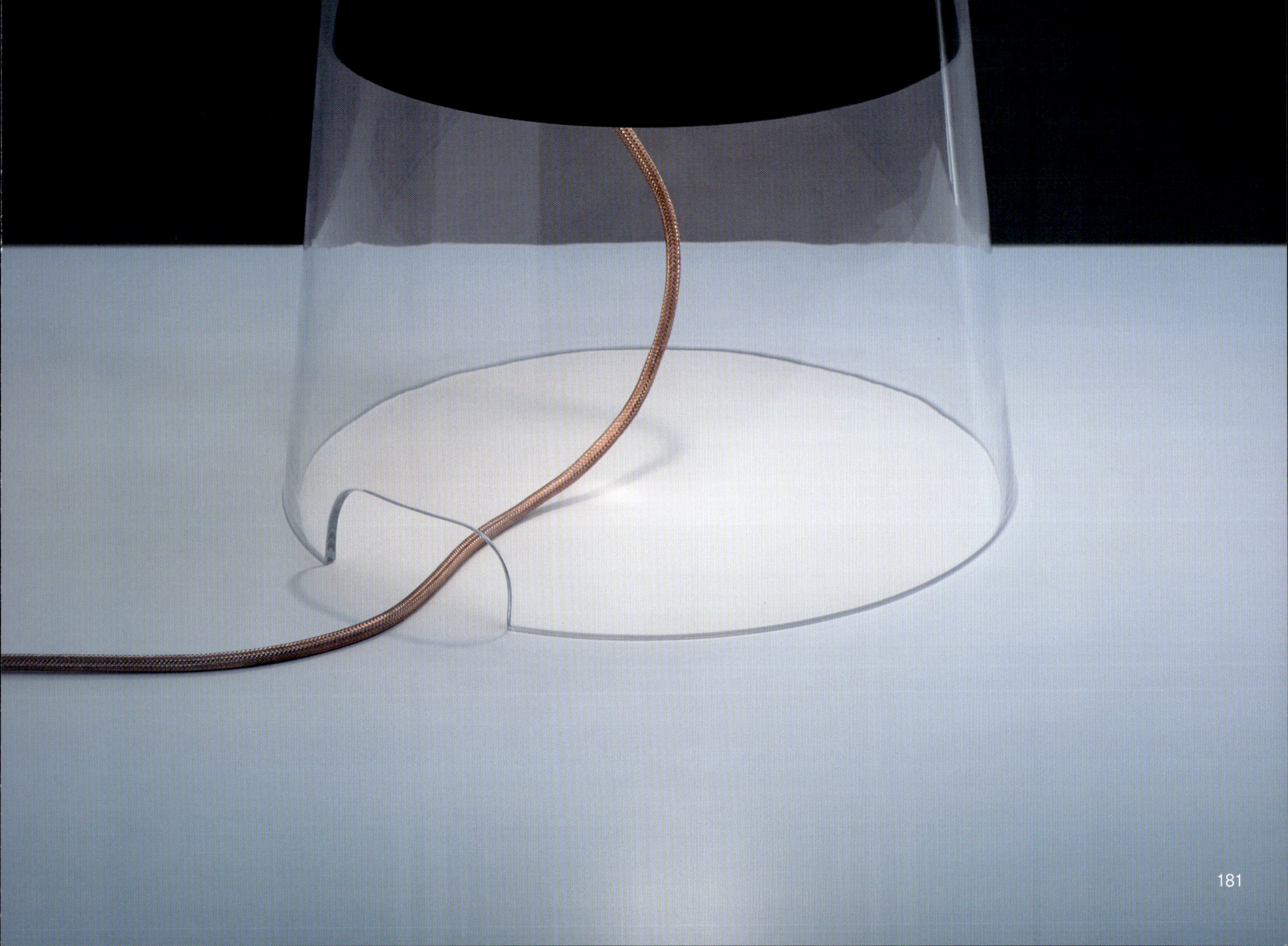

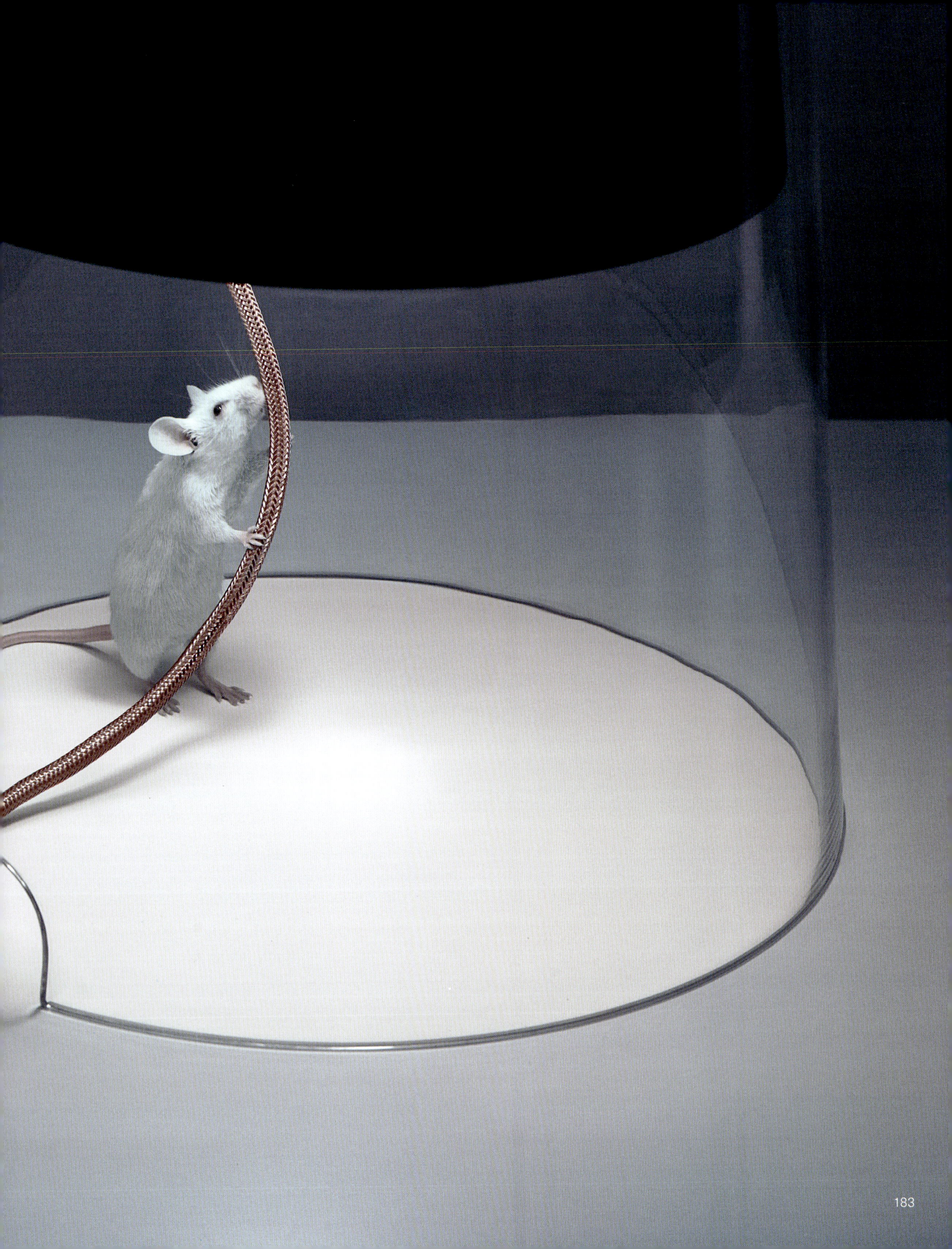

Don Goeman

Don Goeman with Leon Ransmeier

Throughout their history, Herman Miller have managed to attract progressive designers, from far-flung places to a small town in West Michigan in the United States; and Industrial Facility are now a part of this lineage. Lots of people are involved in the creation of each project, yet even within an organization of Herman Miller's size, Don Goeman has a strong presence. An exceptionally tall man, he heads up the Research, Design and Development team that they call 'The Kitchen'. He began his career as an engineer, working with Bill Stumpf and Don Chadwick in 1984, the same year in which Charles and Ray Eames created their last piece – the Three Seat Sofa. Since then, Don has participated in the development of a staggering number and range of products, including many by Sam and Kim. I sat down with Don on a chilly October afternoon in 2016 to discuss his thoughts about these collaborations, what makes a good designer and the complexities of working with Herman Miller.

Leon Ransmeier: Can you remember your first impression of working with Sam and Kim?

Don Goeman: At Herman Miller, it starts with an 'us-getting-to-know-them' phase – it doesn't immediately jump into, 'Oh, we've got a project, and we need this in six months.' I can recall a seriousness and clarity in our first exchanges that was thought-provoking and different to other designers. They didn't use flowery language, but instead communicated in a succinct manner. They helped me to see things that were so simple, you might otherwise have missed them. And then when I saw their work, it was so clean, so logical, that I became a fan from the get-go.

What are Sam and Kim's differences and how do they complement each other?

Well, that's a good question. They're different from each other, first of all. Sam's an industrial designer and Kim studied architecture. I've found Kim's advice to be more big-picture, up at a higher altitude in finding patterns to things. She's closer to research and is sensitive to business. Sam's more human-focused. He's quicker to the essence of things and not always concerned about all the data. Of course, I don't mean any of it at the detriment to either side. They converge in the depth of their work and seem very aligned about their design vision and its principles. I'll ponder it. In fact, I'm meeting with Sam and Kim in London next week.

Well, that's good, because I was going to ask if you travel to London when you're working with them. How would you describe their office?

If I meet with designers in Michigan, it's a meeting squished between other meetings. We're in a Herman Miller space. When I spend time with Sam and Kim in their studio, we're much more likely to have conversations going into some of the contextual places that we never have time for here. It's fun; probably the most rewarding part of my job is getting to know them as professional friends. Their studio is a clear reflection of them in its simplicity and scale. They've got some of their best-loved earlier projects around them.

They're right across from an old pub called The Jerusalem Tavern, and sometimes it's like a part of the shared studio space. It's just a really grounded place. There's no bullshit there – it's all authentic.

You've mentioned seriousness, clarity and authenticity. What do you think makes a good designer?

It's a whole lot of things, to be sure. I don't know that I can fully comprehend all the things that designers bring to the effort.

Conviction for a vision is important, because that gets tested all the way through. To do our best work collectively, we need the designer to be out front pulling us forward and the vision has to be clear. The solution can take many forms, but we've got to be absolutely locked into a notion of what we're aiming for.

I think that it's also important for designers to be great cultural assimilators; and I've always respected that creative people are probably the best at connecting the dots of what and how they see, and what makes it meaningful. It's critical to the Herman Miller formula, because we don't work and live in a hotbed of cultural exchange: we're a bit isolated in the Midwest. Designers like Sam and Kim bring their cultural assimilations with them as a natural part of their work. You could probably name another dozen things, but tying it all together and making something beautiful is an important ability as well – giving form and material a kind of sensitivity. It's one thing to understand the problem, another to frame a vision for a solution; you then have to get something to work in development with Herman Miller, and finally, also make it beautiful. It's not something that just any designer can easily do at the level that we try to practise it.

Herman Miller has evolved quite a bit since the days of Rohde, Nelson, Eames and Girard. The production of systems furniture and ergonomic seating necessitates the need for large teams of researchers and engineers. Would you still describe Herman Miller as a designer-driven company?

Very much so. In the first forty years, the company was dependent on designers bringing everything – their gift set – and, as we have already mentioned, their assimilation of culture. Nelson described design as a response to social change. All of that legacy is still valid and important today.

Probably the most important thing goes all the way back to the beginning. D J De Pree used to describe 'abandonment to the creative view'. That's really tough to do in contemporary business because there's so much data coming out of the marketplace. Herman Miller has always had an uncanny ability to ultimately listen to a creative voice and abandon our internal organizational judgement. I'm not going to say that we don't push and pull things around. If you've experienced working with Herman Miller, you know that it's a mash-up of sorts. But I still think that the most important thing that still drives us today is the creative vision of the designers.

Sam and Kim's designs create a very intuitive experience – they are skilled at making things easy to understand and use. I believe that one reason for this is their strong intuition as designers. I'm curious as to how you would discuss the relationship between a designer's intuition and the need to integrate research into the design process. How do you see that marriage?

Well, it's a little bit awkward. When I think of the research that we do at Herman Miller, it's more applied research. There's integrity in gathering data, and you have to be rigorous about how you interpret it. It's almost more of a science-based approach to understanding phenomena. What's different about designers is that they don't concern themselves with this scientific method. They can see one datapoint and emotionally connect with that as a significant example of something. Creative people see what they realize is important, and they don't much care what the data says. And so there's a tension.

I saw that tension with Sam sometimes when we would meet. Before any design work was carried out between us, Sam and Kim were appointed as advisors to a few execs, the CEO and myself, and this lasted for around three years. At each meeting, we had research presentations from different parts of the company, and I could just tell that Sam was bracing himself because he felt that our researchers were completely missing the essence of something. In many ways, some of the data that gets generated is not that relevant or important to a creative viewpoint, but that's okay. It's all part of the synthesis that happens.

Often, in order to achieve simpler experiences, the process of engineering and developing a design becomes more complicated.

Yeah, there's nowhere to hide! It's difficult from an engineering standpoint – everything is fully exposed. In a way, it's a bit like a building. A modern home costs more even though it has fewer details, because there are no mistakes allowed – everything is visible and open to the eye.

But it's not just in the object or form resolution. When we worked with Sam and Kim on Locale [pp. *218–21*], they discovered through their own design research the phenomenon of people being interactive in one moment among colleagues and the next moment working individually. You could probably characterize this phenomenon with a thousand words, which in the end doesn't get you very far. Sam and Kim reduced the whole project to just four words, 'We, Me, High, Low'. That expression became a driver of our vision. So, it isn't just about driving the form towards simplicity: when they've even synthesized the need for the design to make it so straightforward, you sort of say, 'What? Yeah. Wow. That's it!'

Kim has said that working on Locale brought their office to a greater level of maturity because it was so complex.

It was complex: that's part of the synthesis challenge. In the complexity of a large organization's environment, there are issues of infrastructure, power, data – economic phenomena and preferences and intermediaries in the industry. In office furniture, it's a complex hairball as to how business happens. Still, I think that there's room for abandonment ... okay ... given all that, what does the designer see or think about that nobody else sees or thinks about? Even though they may have had no familiarity with all the complexity way back at the beginning, they'll still point us to a place that is of a new relevance or that's even revolutionary. When I see Sam and Kim, they'll talk about other places that they've been. They keep driving their own learning, and it feeds their notions of contemporary living and working – without the bullshit!

Toilet Brush and Bin pp. 144, 265

Bath Radio pp. 144, 265

OH! Toothbrush pp. 145, 275

Little Disk pp. 145, 263

LaCie

Busby pp. 146, 274

Second Phone pp. 147, 257

1 2 3
4 5 6
7 8 9
* 0 #

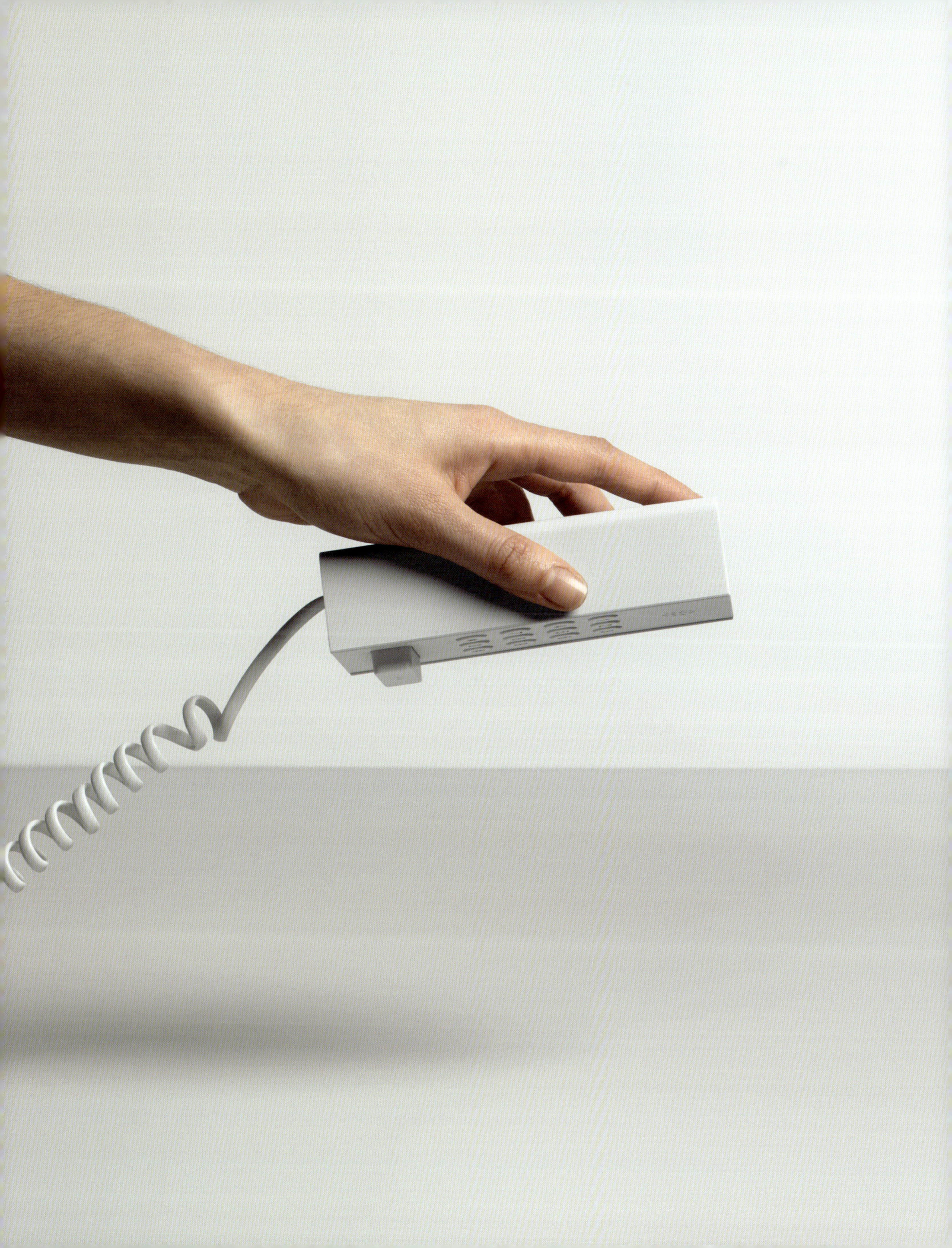

Projector pp. 148, 263

Deyan Sudjic with Oli Stratford

A defining moment in Industrial Facility's career came in 2008 when the studio presented *Some Recent Projects*, a retrospective at London's Design Museum. It was the studio's first exhibition in the UK and was initiated by Deyan Sudjic, the museum's director, whose engagement with the studio reflected a growing critical interest in its work. As former editor of *Domus* in Milan and *Blueprint* in London, as well as director of the 2002 Venice Biennale of Architecture, Sudjic has long been one of design and architecture's sharpest critics and explicators. When I met with him for this conversation, Sudjic had just overseen the opening of the Design Museum's new premises in Kensington, completing a ten-year process to rehouse and expand the museum.

Oli Stratford: How did you first become aware of Industrial Facility and its work?

Deyan Sudjic: I first met Sam when I was editing *Domus* and the magazine covered work that he had designed for Prada during his time at IDEO, working alongside Rem Koolhaas and his practice, the Office for Metropolitan Architecture (OMA). It was for an electronic tagging system that might be commonplace in retailing now, but at the time it was a new development for a company like Prada and their high fashion profile. He presented the project, and the research that had gone into it, in considerable depth. I then started to follow Industrial Facility and was struck by the level of clarity and conviction in their work, which was something that I had not seen before. I wanted to find out more about the studio and the people behind it. I wanted to understand, for instance, the slightly deadpan quality of the projects that they present and I was intrigued to find out whether there is a questioning note to their work, or whether what you see is actually what is intended. I wanted to know what lies beneath that perfectly smooth surface.

What conclusions did you reach? A lot of thought goes into their work, but the studio seems content for its designs to stand on their own, without the need for elaborate explanations.

I think that there are a number of layers that run through their work. They are interested in speculating about the near future, and following this through to often illogical conclusions. Their investigations into the potential of emerging technologies take place in a way that is never wide-eyed, but can be quizzical. I remember the Design Museum commissioned a project on the contemporary meaning of ageing, and they explored what might be described conceptually as the Amazon approach to independent living for the over-eighties. At first glance, it appears an odd approach, but then it quickly starts to make sense. They can sometimes come close to science fiction. But, at heart, they maintain a strong grasp of the consolation of material objects. One particular influence that I see in their work is that of Dieter Rams, a designer who could never be described as sceptical. When we did our Rams exhibition at the Design Museum, I remember Sam looking at the installation with me and saying that it looks like Rams designed everything as if it were meant for one room, which I thought was a remarkable insight. Maybe Industrial Facility does that too. There is a common sensibility that unites everything in its work.

Is that a compliment? I mean the studio prides itself on its projects being contextual and engaging in different situations. It's a practice that seems remarkably devoid of ego or the desire for a singular or 'house' style.

It's the job of any responsible designer to say that they respond to context, production and potential. But if you have limitless possibilities, you can easily become paralyzed and frozen. If something is anything, it's nothing. So you need a frame. Maybe the sensibility they have is that their frame is designed to respond to challenges and different circumstances.

But I get the impression that the studio is sceptical about the idea of having an authorial voice. Even its name is presented as a blank slate: *Industrial Facility*.

The way that designers choose to name themselves is a fascinating, fluctuating wave. Sam and Kim certainly emerged from a period in which the idea of the star designer with an inability to stop talking about themselves was becoming slightly passé. They didn't want to be Philippe Starck (even though they have worked together), and I think that they were probably more interested in Ettore Sottsass, who was also visible as an individual but nonetheless had an ability to make quiet industrial machines with some sense of subdued poetry. Sometimes, it's very difficult to come up with a name and I would imagine that Sam and Kim did agonize a bit. 'Industrial Facility' has got a certain smack of North America about it and the practice is an Anglo-American partnership. I think that they'd like Industrial Facility to be a corporation, in a sense. But I have never asked them why they used that name: it seems to speak for itself.

When I started writing about design, the studio was already well-established, which in part was as a result of journalists and curators like you who knew them much earlier in their career. What was it about Industrial Facility that gave them that position?

They defined a particular part of the aesthetic landscape. At a time of uncertainties – the end of the Memphis moment – they had a sense that they knew where they were going.

But playing devil's advocate, you could say that Industrial Facility's work looks back to a pre-Memphis period. As you mentioned, there is an interest in Dieter Rams and their work is often described as minimalist or functionalist. You could detect a modernist influence in there.

It depends on what we mean by Modernism. If Modernism means believing that a concern for social issues is important, then that's not dead at all: it's a belief that you can use design in a sincere and authentic manner to make life better for people. You could also argue that it's a formal language that came from the Bauhaus and which tried to suggest that it was based on contemporary machine technologies, although it usually wasn't. While Industrial Facility does have a fascination for those refined and simple forms, it is not simply making decorative use of technology. The Branca chair [pp. *62–5*] for Mattiazzi is as far from craftsmanship in the traditional sense as you can get. It's made using very modern, digital cutting techniques to get a shape that is only possible because of them. So, I don't think that the studio is nostalgic, but it does have a hinterland: Industrial Facility doesn't take objects entirely at face value.

In that spirit, one striking aspect of the practice is its ease with words, both in terms of its publications and its general design approach. They begin projects by talking, without necessarily feeling the need to start sketching or iterating straightaway.

Many forms of design practice involve a certain degree of storytelling. Whether it's Philippe Starck telling us that he creates three-legged chairs to make life easier for waiters by giving them fewer trip hazards, or Industrial Facility producing documents and reports that have a faintly impersonal corporate flavour despite their anything-but-corporate content, storytelling is part of what it is to be a designer. But for Sam and Kim to not immediately go for a formal solution does suggest a degree of circumspection about their work – a sense of clarity. Terence Conran always used to say that unless you know how to make something, you can't really say that you've designed it – which may or may not be true. The counterpoint is Vico Magistretti, who always used to say that unless you can describe something simply over the telephone, you haven't finished designing it. Maybe Industrial Facility is closer to Vico than to Terence.

They're certainly fascinated by the notion of the anonymous, unauthored object – something that came to the fore in the studio's presentation of everyday objects in the *Under a Fiver* section of their exhibition *Some Recent Projects*.

They created biographies for all of those objects. They told you what was so interesting about underwear that comes shrink-wrapped in a tiny cube, or a glove that's specially designed for shucking oysters. What you were getting there was an intelligence to analyze an object and tell you why it was special, even though it looked normal. I thought that was really impressive and engaging, because it was analyzing and explaining, and revealing. Maybe design is always about revealing something. You can't say that Industrial Facility designed those objects for under a fiver, but it did reveal them and that's something to do with the studio's design practice. They're revealing an idea or a process through a thing: showing what it means to sharpen a knife, for instance; highlighting the essence of a material; or revealing a possible workplace, as in Locale [pp. *218–21*]. Designers sometimes suggest things that haven't happened yet. They make an idea visible and physical. Industrial Facility certainly does that.

Do you imagine that to be a conscious thing or does it come out accidentally through the process?

I don't think Industrial Facility do accidents. There's something very reassuring about that sense of consideration. If you asked them to design something, you would have a sense of confidence in the result. Looking back on the relationships one has with designers, the ones that work are those where you don't feel the need to ask them to do this, or try that, or could they make it a bit more ... That's usually a disastrous approach and you need a sense that the designer will do something that they believe is right for the problem. But let's not forget the term 'beauty', which is something that people are embarrassed to use or talk about. These are beautiful objects, although they don't go for the kind of ostentation that makes a fetish from the tactility of perfection. Industrial Facility doesn't do that. There's something quite humble about the studio's objects. They don't swagger.

In part, however, the character of those objects is a reflection of the clients that the studio has.

Most designers struggle. This is a time when a lot of people are emerging from design schools, but the number of people who can support themselves through design (as it is traditionally understood) is severely limited. Compare the current situation to Olivetti in its heyday. That company gave Sottsass and Mario Bellini space to maintain studios outside their work as consultants for Olivetti, which provided enormous cross-fertilization and meant that the designers could maintain a very strong, independent position. That's not true of Samsung or Apple today. To work with them, a designer has access to enormous resources and has the potential to make a significant impact, a privilege for which they must trade their independence, however. Most technological impacts have an equal and opposite reaction: a laptop makes it possible to do almost anything as an individual almost anywhere in the world, but to design one of them, you've more or less got to be in Silicon Valley, or its Asian equivalents. If you want to have an impact, can you do that outside of the embrace of a major corporation? Dieter Rams was able to do the 'one room' thing because he only worked for Braun, whereas what Sam and Kim have done is to create that effect with multiple clients, which is perhaps even more difficult. They have produced a body of work that appears coherent and hangs together, without merely reducing that signature to a stylistic gesture.

Store pp. 149, 276

Run shelves pp. 149, 274

Run table pp. 149, 274

Two Timer pp. 151, 265

Established & SONS

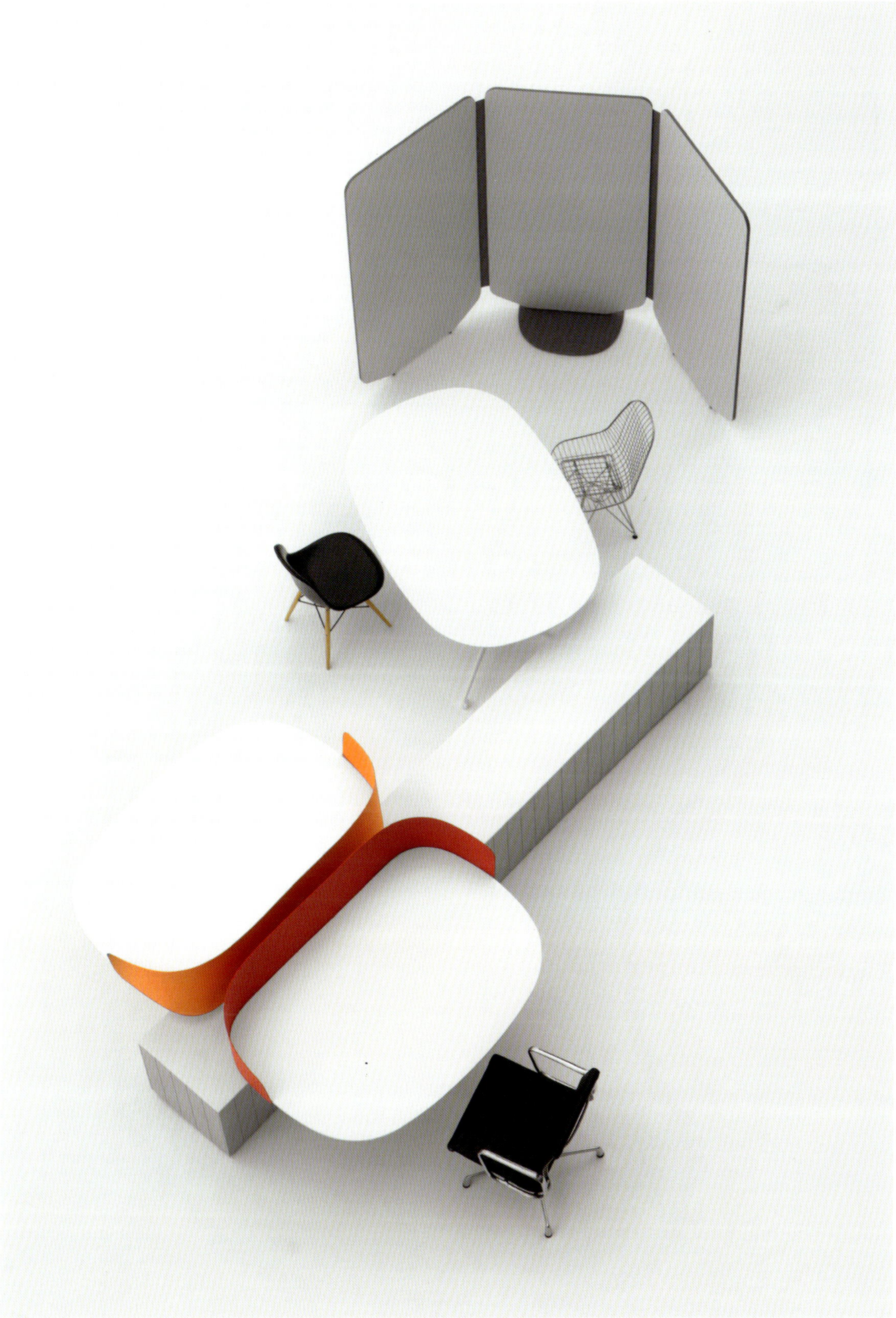

Locale office system pp. 151, 271

Philipp von Lintel and Ippei Matsumoto with Oli Stratford

When Kim Colin and Sam Hecht founded Industrial Facility in 2002, they selected a purposefully blank name. Stripping away any suggestion of vanity from even its title, the studio would become a space in which they could develop projects that engaged with real need, without the loss of an individual style or authorial voice. From among their former staff, I managed to sit down with Ippei Matsumoto (who joined in 2003) and Philipp von Lintel (who joined in 2010). Working around a large, single desk with Colin and Hecht, they were part of a studio atmosphere that prizes group discussion, individual responsibility and creative freedom. Their reflections on the studio's methodology provide an alternative viewpoint on the identity of Industrial Facility.

Oli Stratford: How did the office function day to day?

Philipp von Lintel: I guess typically with a small office, everyone did a bit of everything, from answering the telephone through to all aspects of the design process; but, at the same time, we were each brought in with our own specialities. Sam often initiates the project's direction and also does all of the photography; Kim is really good at framing the meaning or essence of the project and is good at writing; and I'm good with the computer: 3D modelling, complex surfaces, rendering, a bit of animation. Coordinating that depends on the project, but at the beginning of the process we all came together in conversation to try to understand the client, what might be good for the product, and so on.

Ippei Matsumoto: I never felt that there was a singular structure as such. The early years were a bit chaotic as there weren't that many clients. Everything was produced by hand in those days, and we were experimenting a lot because we had the time. Now, the studio has a lot of grounding and experience, so it is possible to have many clients at the same time without sacrificing quality. Those early years were a struggle, but at the same time there were so many possibilities. There was always – and still is – a certain freedom. There wasn't a boss type in the office that told you what to do: some big designers don't really listen as they are trapped by a particular style, but Industrial Facility feels 'real'.

Were you all of one mind in approaching projects or were your meetings fractious?

Ippei: I think that Industrial Facility has always been a studio of reasonable and decent people. There isn't crazy for crazy's sake and we liked to keep our feet on the ground. The studio thinks about the future, but not too far ahead. If someone came up with a better idea, you would just accept it – Sam and Kim have always believed that the best idea must stand up by itself: where it came from is not so relevant.

Philipp: Compared to other studios that I've worked in, we took quite some time discussing through conversation before we started drawing. Then, we would put down ideas, design something, and then we would all meet again to look at it. Most of the time, we would find a compromise through more conversation. That's a working method that suits me very well. Nobody would tell me how to do something, so if I wanted to make a model,

I would do it. If I wanted to visit somewhere to understand a material or process as research, I would do it. There was a freedom there to be taken.

How does that lack of a system or structure square with Industrial Facility's focus on research and detail? It seems as if it could be a difficult set-up to manage.

Philipp: They're not working in a systematized way, but Sam and Kim are in the studio most of the time, so there was always somebody to talk to. There's no loud music, we all sat together at a table, and there was an atmosphere in which you felt that you could ask a question if you wanted to. Initially, my idea of an ideal working environment was that I would work with my best friends. I respect Kim and Sam's opinions very much – it's about the work and none of us held back. We're friends, but there's nothing that would prevent us from being honest. That sense of freedom came about gradually as I matured.

Philipp, you initially applied for an internship with Industrial Facility while still a student. What prompted that?

Philipp: I was studying in Germany and felt very attached to the German design of the past, like Braun, but I couldn't find a German studio that still worked like that. So, when I found out about Industrial Facility, I was very drawn to its work. It seemed to really care about products. It didn't want to express itself in the product so strongly and I was very much attracted to that. In contrast, what I had felt from my tutors in Germany was that they wanted to move on from that Germanic design tradition and didn't want to be stigmatized by the Bauhaus or Ulm School of Design (Hochschule für Gestaltung Ulm). I had to search in the library myself to find out about that period, because I felt that there were parts of it that were still so relevant to my generation.

Why do you think Industrial Facility adopted that method of designing that you admired? And how do you safeguard that?

Philipp: Sam is generally a very efficient person: he's efficient in running the studio and in his daily life. Consider how the whole studio looks, for instance. The top floor is all white (including the floor) and is purely functional, so much so that we can photograph products without artificial lighting; and the lower floor has a workshop and spray room because the ceiling height is higher. In general, Sam and Kim want products to speak for themselves, which is something that they found in both historical German design and in early LA Modernism. I remember when Sam brought in a Braun SK4 record player and together we both tried to understand the whole design process behind that particular product. We were curious for a long time as to why this thing looks so pragmatic and progressive, and were surprised to discover that it involved not one but five contributors to its design: Wilhelm Wagenfeld produced the record-player component; Hans Gugelot came up with the idea of using sheet metal and two wooden sides; Otl Aicher designed the graphical scale; Herbert Lindinger was behind the controls; and Dieter Rams came up with the idea to make the cover out of acrylic plastic. It was confirmation to me that the creative act really is a series of careful, pluralistic conversations and decisions at each point in a product's development, which can involve lots of like-minded people: however, without the company directors, Artur and Erwin Braun, such a design remains a mere concept. There was and still is a common interest in the studio about trying to understand

the reasons behind historical products. Industrial Facility have never been frightened of history – that it may make you feel or appear less contemporary.

What were your early impressions of that method of working and the studio?

Philipp: What struck me the most were the conversations that we had. Soon after I joined, I was reading Otl Aicher's *The World as Design* (1991), and Sam said that it's one of his favourite books. We started having these long discussions about what Aicher was saying, and how each of us interpreted it. That was really the first time that I'd been able to have conversations like that. It helped me a lot to understand what I wanted in design. Until you have an opinion or taste about design, it's very difficult to achieve anything. I'd say that being at Industrial Facility has helped me to put together an understanding of what I want.

How has the studio changed over time? They introduced Future Facility in 2016, for instance, which looks at networked products.

Ippei: The clients have influenced the studio a lot. In the beginning, we were mainly designing products, but after a few years, lots of clients from the furniture industry started coming to us. The core design process is the same, but the scale change does influence you. One of the purposes of Future Facility is that the studio wants to look at products again, and without talking about the Internet of Things, it's not really possible to do that.

Philipp: The environment of consumer electronics has changed. Within the studio, there is a healthy balance of being interested in technology coupled with an awareness of the convenience that it can bring, while also being conscious of the danger of it starting to control us. One project that came out of these discussions was the Busby light for Wästberg. It was felt that the reality of daily life had changed (as it continuously does). We need light at the bedside to read, but a lot of people charge their devices there now too. That idea was around for some time in the studio: Sam had first made a model some two years earlier to investigate how power and light could be connected sensibly, and execute a function with some relevance. Kim has no time for gadgets, so she always judged if the concept was fundamental and if its material outcome was right. There were a couple of iterations of the design that failed, but we kept with it. The final result doesn't dramatically change people's lives, but it does respond to change.

How would you go about developing a project like that? It's been well documented that physical models play a big role in Industrial Facility, for instance.

Ippei: Product design is essentially for people to use and to touch: you feel it and touch it, and if you don't also do that in its development, then it doesn't work. We always said that if a product is meant for the hand, then it should be designed by the hand – through the act of making. In order to reach one form, there are often ten or fifteen models produced by us, but rarely seen by the client. That's because the process is quite intimate, in that you're making one mistake after another. After each model is made, there is a conversation and sometimes a pause before making another. You don't want a client to witness this process, even though they often ask to. It's a bit like being a chef – the last thing that you want is the customer tasting the food as you're preparing it.

Philipp: People think that Industrial Facility must be quite a big studio because there are so many projects in production, and they're quite shocked to discover that it's always a small team. But I think that people admire the consistency of the output. Every project has its purpose and its place. People have often come to me wanting to get hold of something that the studio has designed, which is nice. The desire is to make things that are relevant and that people are happy to own, which is why it's called Industrial Facility. It's really less about the studio than the product. It's good to get clients and be known, but not to strive for that. It's a belief in people having things around that they want to keep.

Locale whiteboard pp. 151, 271

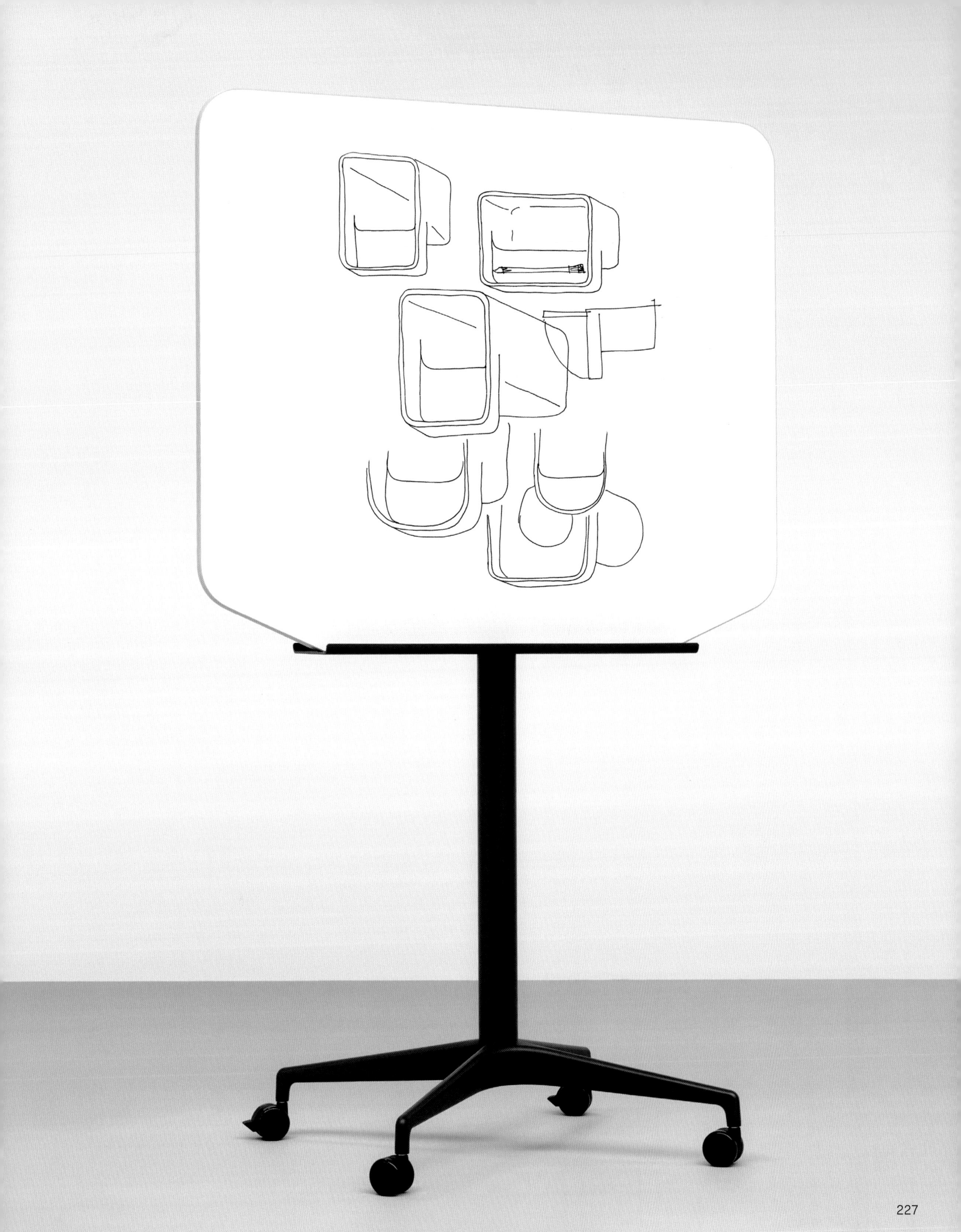

Formwork pp. 153, 273

Locale table pp. 151, 271

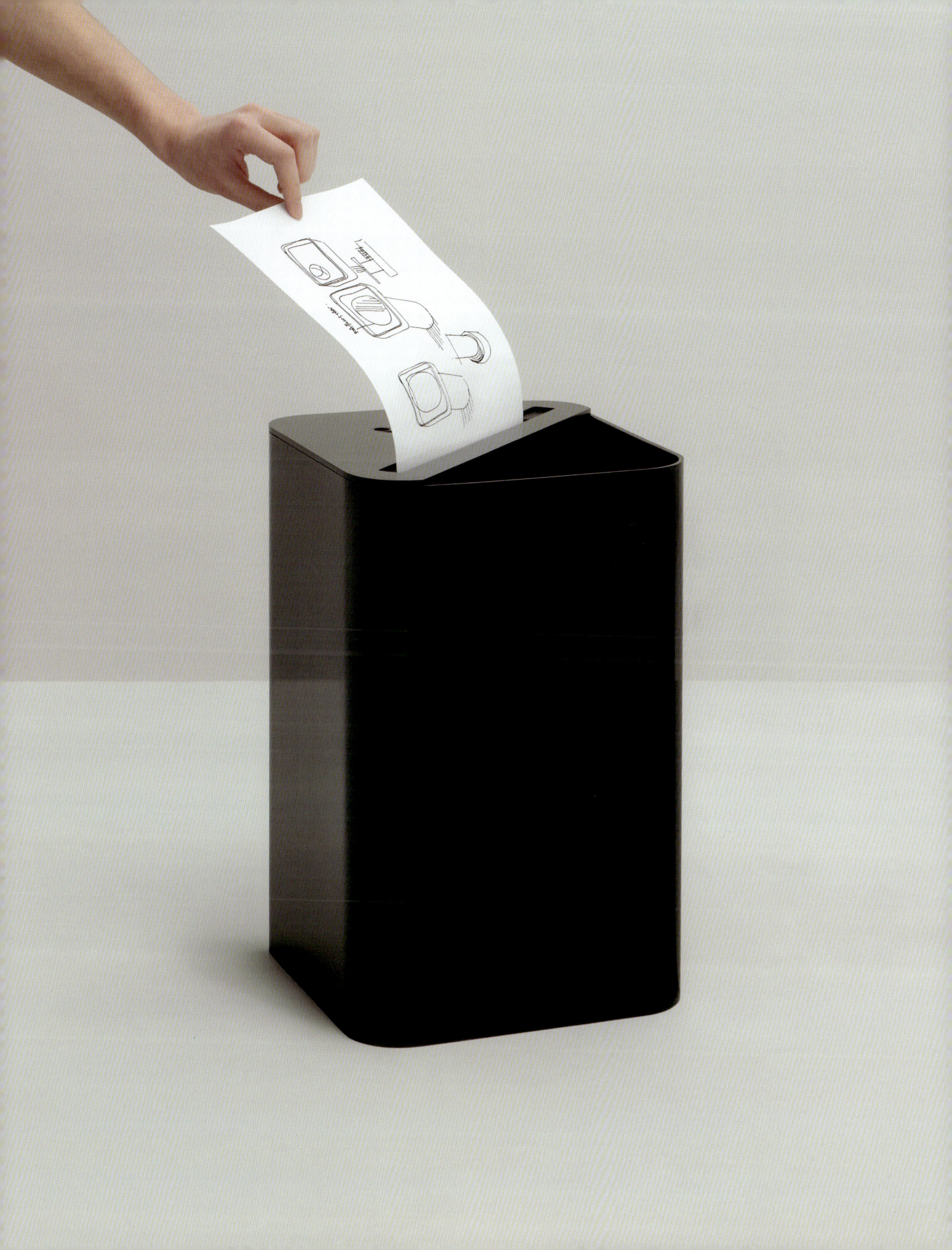

Shredder pp. 154, 265

The Origins and Working Methods of Industrial Facility

Lecture delivered by Sam Hecht in March 2016 to an audience at Kyoto Institute of Technology, Japan

First of all, I would like to thank Sosuke Nakabo for his introduction and also to Julia Cassim and the Kyoto Institute of Technology for inviting me to conduct a workshop on design and to give this lecture. I am going to describe our studio in London and the mishmash of different ideas that we follow. We have no real set pattern or process; no methodology as such. The majority of the work that we do is generated from conversations within the studio, between us, and with the many factories that we work with, our clients and their customers. Nevertheless, there are strands to how we think, which I will attempt to describe. All of these strands converge in an effort to make design a provider of enjoyment, pleasure and relevant things, so that we're happy for manufactured goods to stick around and not simply be destroyed out of frustration or thrown away after a short period.

I started our office with my partner, Kim Colin. I'm English (I was born in London), and Kim is from Los Angeles, so she is from a very different continent, a different place, and her background is different too. She is an architect by training and never studied design, but art history. We came together through conversations. She was teaching at the time at the Architectural Association in London and I was very curious about architecture, and always have been. But I was more curious about the process of how architects approach their work and whether it is feasible to adopt such a process in design. After all, the responsibilities of a building can far outweigh that of a product or a piece of furniture; but I felt that it was the right time to take up a greater responsibility for the things that we use and that surround us, and so I wanted to learn from Kim.

We decided to form a studio together in 2002, after both having already developed careers that we would leave behind. As Sosuke Nakabo has mentioned, both of us had already worked for small creative studios and large corporations: it was vital to experience such infrastructures. Our experience informed us that the scale of a company can denote the roles and responsibilities of the individual: we were adamant that we wanted to try to do all of the work ourselves and that there is no place to allow this other than a small studio. We knew that our office would be an intimate place and it still is. We are always a group of around five people, which is quite a common number in design. Yet, a small studio will always have its limitations in the work that it can do. We wanted there to be no limitations in the scale or complexity of projects. So, to overcome this dilemma, we deliberately chose a studio name, Industrial Facility, that would confuse people into imagining that we were much, much bigger than we were – as big as the corporations that we had left behind. If you think about it, the name Industrial Facility doesn't mean very much; it is meaningless (even though, over the years, we find ourselves post-rationalizing explanations), and it is open to interpretation. When the phone rings, people often ask for different departments. There is only one. We did this because we knew that it would be a potential way to be able to talk to larger industries regardless of the size we are.

In the office, we also have design assistants from very different places and backgrounds. They can be from Germany, Italy, the USA, Japan, Spain, France and so on. So, it is an extraordinarily international group, which only a city like London makes possible.

Now, I'm going to talk about these different strands of thought that continually run through our office.

The Importance of Context – More than Just Product

Design should be invisible – it should not get in the way of its use, but rather, it should make sure that it is pleasurable in use, from the placement in a room to its relationship to what surrounds it.

The first strand is what we call 'more than just product'. When we work with a company and we're developing a design, we rarely think about the product directly. We're thinking about what's around it. In other words, we go much further than the brief, regardless of whether the client is interested in us doing that or not, because that is what will give the project its grounding, its purpose.

To appreciate this grounding, I will give an example. If we look at an image of a glass of water on a white background, everyone can see it, everyone can understand it as a glass of water, but the reality is that this is simply an image. It is not possible to see this in reality, because this glass of water or the glass by itself will always be connected to some kind of context – and cannot be removed from it. The glass could be on a table, or in someone's hand or on a shelf. It could sit on a bedside table, or on a working desk, or be moved using a tray. So it is important to appreciate that if you only design and consider the product – the glass – then you will end up with just a product, without the reality of how it spends its life. But if you contemplate everything around it, its context, then you'll end up with something much more. This is what we mean by more than just the product.

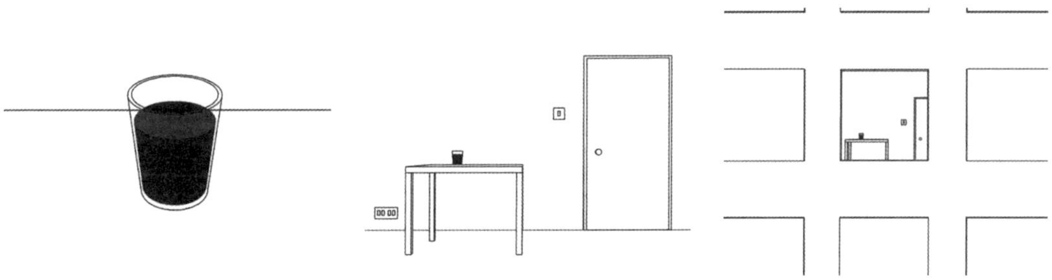

For us, it's almost as if when you see this glass, you have to imagine that it's connected to the table and that the table is in a room, and that the room is in a building. In other words, there is a connection to all of these things that's impossible to ignore and, in fact, you shouldn't. These connections are what help to make sure that ideas are relevant. Ideas can sometimes feel obvious, restrained or a little boring, and there can be an urge to invent things that are not there. Sometimes, what is seen to be exciting can also lead to short-term relevance. A product is not a TV show – it cannot be turned off. It is physical and must live alongside people who are sometimes happy or sad.

To try to explain this essence of context even more, we often use the example of language. Here is a wonderful example of a paragraph of text that I can read very, very easily, even though it's jumbled up.

> Recently, I got this strange email. I cdnuolt blveiee taht I cluod aulaclty uesdnatnrd waht I was rdgnieg. Apprantely, it dnsoe't mttaer in waht oredr the ltteers in a wrod are, the olny iprmoatnt tihng is taht the frist and lsat ltteer be in the rghit pclae. The rset can be a toatl mses and you can sitll raed it wouthit a porbelm.

It appears visually as a mess with lots of spelling mistakes, but is at the same time entirely legible. We are taught that correct spelling ensures correct communication, but as you can see in this example, I have no problem reading it – even though theoretically, I shouldn't be able to! The reason we can read it is that we don't read individual letters, we read words, sentences and paragraphs. The letter by itself is meaningless, but put together with other letters it forms a word that provides meaning. And so, even though the letters might be jumbled up, our brain is easily able to reform them because we're searching for words; we're searching for the larger context beyond the letter.

In many ways, the glass is similar to a letter, and design should not concern itself only with designing this letter. We are designing the word, the sentence and the paragraph. This is what our ambition is: to not feel success in the product or the letter itself, but in how well it can sit with other letters to form a language that gives meaning.

When you translate this thought into products that we've designed, one example is the Coffee Maker, designed for Muji [pp. *72–3*], which was produced in 2006. The idea was to create a coffee maker that has a relationship to the place where it is going to live. At the time, the majority of coffee makers from a top, plan view tended to look mechanical, where you have a water reservoir at the back and then the coffee filtration at the front. What we wanted to do with Muji was to simply create a circle. Why? Because a circle is a wonderful form in domesticity: there is no beginning and there is no end, and a product with no orientation. So, it can sit in a corner or it can sit in the centre of a counter and look very natural in a kitchen setting. Already, cups, pots, pans and plates are circles. We're thinking about the context; about what's around it.

The Micro Versus the Macro View

Design should be like a friend: we should be happy for it to stick around, to introduce it to people and to take pleasure in its company.

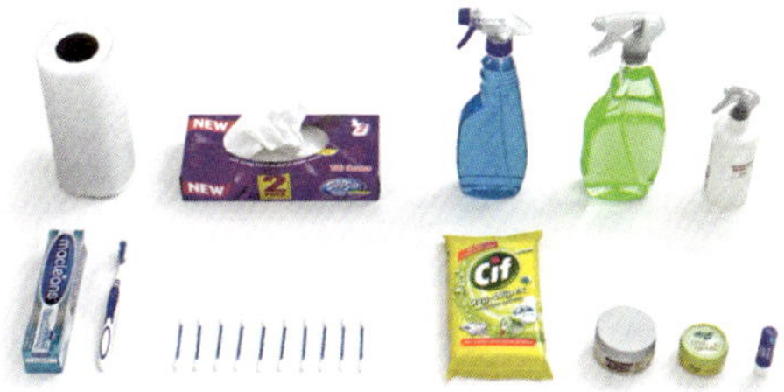

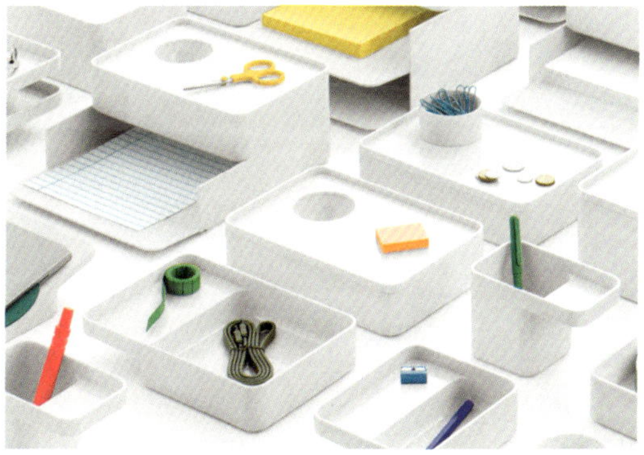

Another example of the importance of context is Formwork [pp.*228–9*], a project that we did for Herman Miller in the United States. It looks at storage for the desk or around the desk, for all of the little things that we can't seem to live without. Again, what we're doing is looking at the reality of the things that exist in this world to try to acknowledge it a bit more. I have seen many designers take pictures of lots of people's desks at work and they assume that they will be able to glean some information that will then influence their work. But this is not the type of investigation that we do. What we are doing in our studio is entering the reality, the micro-reality, by looking at the individual things that exist, not only the photograph from afar, which is a macro-view, where all you see is natural mess. Instead, by examining and acknowledging the micro-view, you can start to have a proper design conversation. You start to see cleaning fluids, personal care items, food, stationery, books, keys – in fact, an amalgamation of all facets of daily life – played out on a single desk. If you concern yourself only with the macro-view, then you see mess and you believe that your job is to clean it up – to beautify it – which is not an acknowledgement of real life. Here, beautification is not going to get you anywhere: organization requires an understanding of the micro-view if there is to be any hope of making a significant impact.

Why Things Are the Way they Are

The next strand is what we call 'why things are the way they are' – a determination that we have to investigate reality in use and manufacture; investigating why things are the way they are, because only then can you actually design. It's very hard to design at arm's length: you almost have to touch it. And that's what we're always trying to do, to pursue knowledge. This is what the independent designer is obliged to do: to gain knowledge constantly for each project. The film *Cast Away* (2000), directed by Robert Zemeckis, is one of my favourite films. Tom Hanks plays someone who has been washed up on a deserted island and does not have the knowledge to make fire, the most basic, primordial activity of humans. When he is able to finally do it, it's a miracle – something unimaginable to him. Making fire is essential knowledge and this is what we're constantly searching for – the basic knowledge, and not necessarily the complex knowledge. I will now describe our first furniture project, which was developed for Mattiazzi in Italy, not far from Venice in the Udine region, which is known by generations of designers for wood furniture production. They know everything about wood and how to shape it, and have very advanced methods, as well as great craftsmanship, within the factory. In other words, they have knowledge that we do not possess. So, it was a process for us to be able to investigate how they produce furniture; and, only when we understood, were we able to create the Branca chair and table, in this example [pp. *62–5*].

Once we analyzed and gained the knowledge, we were able to design a chair to maximize what Mattiazzi is great at. Normally, a handcrafted chair of this type could take a week to make by a craftsman, but in this case, by utilizing a combination of advanced 5-axis CNC capabilities alongside the craftsman, Branca can be produced in less than one day. It means that such a chair becomes more accessible – beauty becomes more accessible. It is still special, but a little more affordable.

Simple Hides Little

Design shouldn't be viewed as temporary. We should all have a feeling of angst if an object breaks or if a part wears out, and doggedly try to save it.

The next strand is 'simple hides little'. I would be the first to admit that we do like simple things. The reality of this trajectory is that you can't hide behind simplicity, as all is laid out before you. You are forced to find the essence of the object or the product, and it has to live by itself through its simplicity. Without this essence, it will not survive for long. Often, we are faced by products in the world that are confusing, complicated and have features that we don't need; that get in the way of this essence. They are produced from situations where companies ask customers what they would like. And the struggle with this process is that while it would appear good practice, customers will always say,

'we would like more'. No customer will say, 'I would like less please'. Customers would like more of this and more of that. But the problem is that more doesn't necessarily mean better. In fact, when more is given, we find ourselves yearning for the time when there was less – when things were simpler.

But it would be foolish of me to suggest that the process of design itself is anything but complicated. The route to simplicity is often chaotic, spiralling and rebounding on itself. Simplicity is not a trend or style: it cannot be applied but only worked towards. It is an arduous and time-consuming process, in which things are made easier and more enjoyable to use. The investment of resources and the care necessary to achieve simplicity are so great that few products reach this standard.

Character or the Essence of the Design

Design is radical only because of its uncompromising essence: its form and communication are relevant to the life of the product and not only to the shop.

A further strand is the idea of character, which is very important. Even if things are simple, they need to hold some character, but the character that we are interested in is where it is born from the idea; the actual essence of the design, rather than applied to the idea arbitrarily. As an example, this is a lamp designed for the Swedish company, Wästberg. It's an in-between product, in that it deals with power and light. When it was photographed, the photographer had a baby with her, and the room was full of things, visually filling up your peripheral vision in every direction. For some reason, the baby was immediately attracted to this little product, called Busby [pp. *198–9*]. Even though it doesn't have a particularly strong character, meaning that it's not particularly expressive, the baby was only interested in this lamp. Something so simple can still hold quite a bit of character.

What we always enjoy, although it can make the project difficult for us, is to challenge the company and the territory they are occupying. For example, Emeco had not produced tables before approaching us; they only made chairs. But we wanted to bring to them what a table would be if Emeco made it: a place for all their chairs. Similarly, with Wästberg; they only make lighting, but for us, it seemed perfectly reasonable to imagine power with light, in this case, USB power, which has become a necessary power source. This was not some added feature or afterthought, but central to its use and to its character. Busby can charge the latest laptop computers, where you don't need mains power anymore.

Locale

Design can respect standardized industrial methods, yet at the same time make the object and material worth more.

Lastly, I would like to share with you one of our longest projects, the office furniture system for Herman Miller, called Locale [pp.*218–21*]. The thing about creating a furniture system is that the word 'system' means that it can never stop. You're always trying to adapt it and perfect it. It really is more like an ecosystem. The lovely thing about Locale is that it's trying to soften the office dramatically, while giving an enormous amount of adjustment and flexibility. Even though it's a very large system of components and products, it is visually simple. What we noticed was that in a typical office, there are round tables for

meetings and rectangular desks to work on. The thing about the rectangular desk is that it immediately has an orientation that is closed for other people to join. It means that it's fine for an individual, but as soon as a colleague tries to come to your desk to have a conversation or collaborate on a project, it becomes hierarchical. It's often at a low height, you've got to look for a chair and it feels like you are at someone else's desk. But a round table doesn't have a beginning or an end and you can enter it from any point. So, what we're attempting to do with Locale is to take the roundness of the meeting table and to say, 'why can't your desk be a meeting table as well, so that you can actually collaborate and work with people, and not just go to a meeting room to do this.' Not only are the desks rounded – there are no legs! The tables are cantilevered and height-adjustable, which means that there are no obstacles for coming together and working. It's proven to be popular for people who use it and we continue to add new products. Once a system is created, it will continually need to be perfected and honed.

As you have witnessed, I have attempted to explain how we work, and what we are moved by. But some of you – the curious ones among you – will walk away a bit confused. Confused, because some of these strands of thinking are in contradiction to one another. How can you talk of simplicity but also of character? How can a micro-view of the world lead you to a macro-view? I would be the first to admit that it is a mishmash. And here lies the realization that design, as a subject, does not have a right way or a wrong way. It cannot have a prescribed process because the route to a result is by nature a creative act. To make things even more confusing, I will leave you with some more contradictions: that there is complexity in purity; that there is elegance in plainness; richness in reduction; depth in the minimal; surprise in uniformity; and innovation in reuse.

Thank you.

Beauty as Unfinished Business, 2015

Introduction to the exhibition, *Beauty as Unfinished Business*, held at the Biennale Internationale Design Saint-Étienne, 2015

We do not talk about beauty in the process of designing because it does not drive us. But sometimes other people talk about our work and use this word. Beauty is not a goal, but instead a by-product, ascribed by others.

We have no intentional formula to achieve it, nor do we try. We think of beauty as an immaterial principle, and less as a material result. Even when faced with what could be 'the sculpture of wood' when designing a chair, the space left open to decisions is as wide as possible. Many have remarked that our Branca chair for Mattiazzi is 'beautiful' [pp. *62–5*]. They run the palms of their hands along its arms and down its legs, marvelling at the seamless wood. Yet, this wood, and its diameter and form, were appropriated from a broom in our workshop.

Do people run their hands down the length of a wooden broom and feel touched by beauty? So, we ask, when do we come in contact with this quality that is untouched by human thought; and can objects ever really reach it? Beauty in products, in our opinion, has to do with the unity of material and function. However, there is a spatial dimension that is also very necessary and rarely ever addressed: the product must be experienced within a context that arouses something in us. This 'something' happens when we look beyond the thing itself, beyond its own space, to something much bigger than ourselves, without time. Without space around it, the object becomes 'flat'. Context adds an atmospheric dimension of space and time, while the tension or energy between the product and its context makes our minds and souls struggle just enough to raise questions. This is perhaps as close as we could come to describing beauty.

How to Design a Chair Sam Hecht
First published in *Monocle* magazine, April 2014

When designing a practical chair or a stool, there are two important considerations: comfort for the body, and comfort to the eye. Both need to be designed, but there are no particular rules that need to be rigidly adhered to. This is why there continue to be hundreds of new chairs created every year, because the chair is the canvas with which a designer can express themselves most effectively. Imagine a pendulum that swings between these two forms of comfort – the eye and the body. I like to position myself just to one side, not forgetting the other side, yet also not looking for perfect balance. Tension is what creates beauty. A simple example is a broomstick. To find the perfect curve that matches this item, try to bend the stick as hard as you can and – just before it is about to snap – this is, in my eyes, the perfect curve. You can see this trick in the Branca chair [pp. *62–5*] or the Radice stool [pp. *80–3*]. Even though Branca is not bent into shape, its drawn curves have come not from a computer, but from a wooden stick that has been bent under tension. This structural tension can apply itself to any material. In aluminium, the Landi Chair (1938) by Hans Coray holds this exquisite tension in its seat, where applied pressure stretches the material until it can stretch no more. In plywood, Arne Jacobsen's Series 7 chair (1955) appears effortless in its combination of rigidity for support and flexibility for comfort. The tension is provided by removing as much material from the structure as possible, to the point that any more removal would render it useless. This is why we subconsciously find it so beautiful. It is why a copy of this chair fails to share the beauty of the original, because it has lost its tension, mostly by playing it too safe.

Reality Rather than Image
First published in *Locale*, 2014

For many people, buying a product is often the method by which we are able to secure the belief that our needs will be fulfilled. It's why an enormous amount of energy is spent on the act of persuasion to buy the product. The product is the 'easy' reality that can be seen, touched and operated.

In truth, though, people don't buy products even though they think they do. Instead, people buy a part of a system, and it's the product that often acts as the simplest and most rudimentary way to represent this system. It's not the camera or the TV that we are buying – it's the system behind the product that lets us do what we really intend to do, to watch TV programmes that we like. To design a better TV remote control is not to make it bigger so that more buttons can be accommodated, or to make it more ergonomic in the hand. A better remote control firstly acknowledges the essential point of the product – in this case, to control and navigate media – and makes it easier and more graceful to us. Of course, when you don't own the system, it can be hard to innovate in the essential territory that needs to be innovated; but it's striking how, even when companies do own the system, they still try to hide it – maintaining the illusion that it's all about the product and not the system.

Recently, this status quo has started to change. Some designers and engineers have regained the territory of the system, and have shown that by visualizing it in a straightforward manner, consumers realize just how much easier it is to control. The Internet of Things is a good example, because it shows consumers that a product is not necessarily the point of what you are buying. A thermostat produced by the company Nest may appear like a product, with smooth lines, packaging and an instruction manual, but what it actually offers is a gateway to greater and simpler control of the temperature in the home. The real product is 'temperature control', which can be accessed in a multitude of ways, learning your preferences as time goes by, to the point at which the product you originally bought in Home Depot literally disappears.

It takes great courage to follow this pattern of thinking, but the results can be striking. With just two products to its name (and after only five years of existence), Nest was bought by Google for 3.2 billion dollars. It wasn't the product that was of value but the system behind it, which was visualized so well for the customer to control effectively.

A designer, and to some extent anyone in a field of development, needs to be naturally curious. This is because necessity drives people to seek common-sense solutions to their problems. Real life – rather than an impression of what it might be like – is the designer's playground. A lever-arch file was never intended as a method to keep a window open, but it is highly effective and always at hand. As Douglas Adams noted in his 1992 novel *Mostly Harmless*, 'a common mistake that people make when trying to design something completely foolproof is to underestimate the ingenuity of complete fools.'

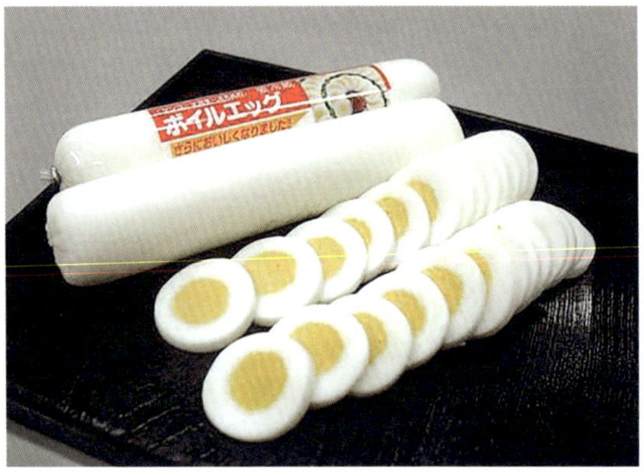

Curiosity is unfortunately not something that you can teach – at least, as teachers, we have never found a way to instill it into a student if they are not already curious about the world and 'why things are the way they are'. Curiosity is the fuel that design depends on. Not being content with what is in front of you is very much the instigation for discovering the reality. Why are paving blocks the size that they are? Because they were designed to match what labourers could physically handle and position into place. Why do some old houses have bricked-up windows in Europe? Because there was once a 'window tax' charged on each property, dependent on how many windows it had. Why do boxed egg salads rarely contain the end of the eggs but only the sliced egg from the middle? It's because, as designer Oscar Lhermitte found, producers are using what is called a 'long egg', an extruded and processed egg sausage that is sliced with no wastage. In the manufactured world, things are continually altered, yet we are often unaware of the manipulations required to fulfil our expectations.

Getting under the skin of the 'image' with which we see can reveal the reality of the system that created it. Only after thoroughly understanding the realities of how products came to be, their system of production and control, can the activity of design take place. When it comes to products, furniture and the manufactured world, curiosity and design are inseparable, because they are representations of systems.

Design is Much More than Product
First published in *Locale*, 2014

The products that inhabit our lives are likely an assemblage of oddities. What we may buy for ourselves is often mixed with mementos (things that we find or have inherited), shared items or gifts from others. Products are sometimes not even owned by us, but hired, contracted or are free with a subscription. Even though living with products is the result of an accumulation of time, taste and sentiment, it is certainly fuelled by industry, whose general goal is to sell as many things as possible. (One of the key national economic indicators is the manufacture of consumer goods for both domestic use and export.)

As consumers, we have increasingly come to realize that this 'selling the most possible' model is a threat to the world's limited resources and the result of rather short-term thinking. In fact, the word 'consumption' actually means 'the using up of goods and services with an exchangeable value'. Using up! Industry knows that consumers are becoming leery about their intentions, so the answer to increased public consciousness is to turn to what is called 'sustainability', the idea that we can go on sustaining our consumption levels, because products or materials can be recycled in an afterlife, once they have become either functionally or aesthetically irrelevant. But is this really a good solution?

Of course, there is another type of sustainability (that we think is underrated), and this lies instead in ensuring that products have qualities of longevity and endurance that justify their manufacture and distribution. We know that with the right balance of quality and a character that we enjoy, we are happy for products to stick around for longer, when given the chance. In the end, this is a far greener and more efficient form of sustainability because, chances are, if we *do* grow tired of the thing, there's a strong likelihood that someone else might appreciate it or find it useful after us. What is the point of designing for sustainability if the thing has no enduring quality, compared to designing something that's not necessarily designed with the sustainability moniker, which will stick around for much longer? E-commerce retailer, eBay could be regarded as a great recycling concept for our times, precisely because it provides a marketplace for the second life of worthy products.

Part of the responsibility of a company is to act as a proponent of the idea that good things should last a long time. There should be a heritage of investing in design; demanding a quality of idea that is beyond material, so that the things that are produced have longevity beyond fashion, in order to keep them in circulation. An Eames fibreglass chair might have been appreciated for fifty years by someone and, in all likelihood, will last another fifty.

Part of the responsibility of a city is to ensure that the communication of design is necessary. Are three cycle signs on a road necessary, when one will do? On the face of it, open space sometimes becomes the most simple and enjoyable characteristic of a city – an antidote to overwhelming noise and brand.

The desire for endurance to be a requirement for products demonstrates an understanding that good design is about equilibrium: the balance of many forces at work when commerce and culture mix. Ambition, material quality, inventiveness, price, market size and so on – all of these forces need to be balanced in an equilibrium, where the final product feels 'just right' and justifies investment in design, manufacture, purchase and distribution (and everything in between).

Dependence vs Renaissance
First published in *Product as Landscape*, 2005

Products rely on things other than themselves to exist. Built into this is an idea that we call dependence – that an object depends on something outside of itself for resonance and to complete the picture. As a designer, you can appreciate the purpose of design more when you admit that products rely on other things to exist. Manufactured things are often there to serve something else. By avoiding the belief that a product 'can do it all', you can acknowledge that the people and a room participate in the purpose of a product. Might this equation potentially mean a different kind of telephone? A different alarm clock? We have used this argument for evacuating functions from objects (that are usually overloaded), because something nearby is already doing it. When we see a product or a piece of furniture that expresses itself so strongly and so resolutely that it renders what's around it insignificant, it turns the room into a shouting match, and design has done a disservice.

Product as landscape is a reflection of our fascination with what binds material to the concept of pure necessity. Physical landscapes have always inspired us, as seen in our images of accidental situations or in the mesmerizing landscapes of faraway places. Somewhere, deep in all of our work, is the idea that products require landscape in order to function. These products deliberately come with less, because they are supported by something much bigger. If we acknowledge that our lives require an ever-greater capacity to sustain overlapping functions, then landscape plays an even more important role in harmonizing our world.

For a long time, the things that surround us have served our needs fairly well. Each encountered problem was serviced by specific product solutions. The premise was always that the human was at the centre of consumption, with all things directed towards fulfilling a perceived need: every object's raison d'être was a result of human impetus. But now, the old path of problem-to-solution/complexity-to-simplicity seems less relevant. Schools teach less about problem-solving and more about opportunism. Developed skills have been superseded by the importance of social alliances. We find ourselves increasingly in a state of problem-less cycles.

Have we run out of problems? Instead of materialistic instability, we now have political and social instability. Instead of being driven by desire, we search for reduction. We now accept that products and systems have their own infrastructures. Objects talk among themselves and, in truth, people are but a part of the balance of life, rather than its focus. So it seems pertinent to think of products not in the singular, but in the plural; sensitive to proximity and to how things lie next to each other.

The Beauty of Equilibrium

First published in *Locale*, 2014

In physics, equilibrium is 'a state of rest in which there is no resultant force on a body'. In chemistry, it's 'the condition existing when a chemical reaction and its reverse reaction take place at equal rates'. In general terms, equilibrium is 'a state of balance due to the equal action of opposing forces'.

The naturalness of all things natural could be said to be the result of them reaching a state of beautiful equilibrium. If we were to take a bottle of water from a refrigerator – a machine – and rest it on a table, it will, over a short period, adjust itself to match the temperature of the room it is in. The water finds equilibrium whether it is in the room or in the refrigerator.

In design, there is often a presumption that to achieve naturalness in a manufactured product is to imagine it in a singular environment, when in fact it might migrate in unimagined ways. The designer, through no fault of their own, imagines an ideal room or a situation that the product or piece of furniture will occupy. How often have we tried on a dress or a suit in a shop and felt visually or physically comfortable with it, only to find that it changes its appearance when we take it home? We may start to have second thoughts, even though the physical clothing has not changed. What *has* changed? The light that surrounds us; the colours; the proportions and the materials of the room. In essence, the atmosphere has changed and so has our relationship to it. We have bought the clothing in a single context, even though in reality it will be worn in all sorts of conditions. The same can be true with office furniture, where the showroom is a perfect representation of a particular context, or a certain way of operating. It is why interior designers, when planning an office, must also deeply understand the company culture, the architectural setting and the people who will use it in order to reach an equilibrium of style and use in the real world. But equilibrium is not the same as neutrality – because matter itself must change according to the environment it is in. A fixed, bland and uncharacterful furniture concept (which is the price paid for greater general acceptance) can sometimes deplete the underlying ambition of a potentially joyful and productive workplace.

Why does everything look so beautiful in the Eames House? Why is it that we have an urge to consume in some shops more than others? What is it about the warmth of a Seattle coffee house or the endearing simplicity of an authentic Japanese restaurant? Perhaps it's because of the complete environment, not just the things themselves, that connects us to the idea of an equal beauty. These special environments create a landscape that makes objects a conduit to a form of balance, so that we sometimes call them 'beautiful'. The context adds something that the object can never hold alone: the relationship between the thing and its space is everything.

With mass-produced products, some designers want to achieve equilibrium of form, but without considering purpose. Similarly, they may want beauty of function, but without an appropriate aesthetic in the form. Between these two positions is another perspective that somehow acknowledges the importance of both aspects, where what a thing does is as beautiful as how it looks; where part of its function is how it looks. We are therefore not describing beauty that can be dissected by extracting one quality of the object (its surface, colour, form, innovation, materiality, etc.) at the expense of the thing's other

properties. With something beautiful, it should be impossible to discuss one of its properties without relating it to the whole of its purpose. While some companies seem to have paved the way for wanting to manufacture more beautiful products that reach a state of equilibrium, we believe that the product cannot be successful by manufacturing a purely aesthetic condition.

In the development of products, there are many forces at work, just as in nature: there is a wish for the functional performance to be better than before; the cost of how much it will be to make it and how much the market will bear to pay for it; the time that it takes to engineer and complete manufacturing; the overall appearance and character of the object; how it will be distributed; and, of course, the whims and opinions of all of the company executives who must bear the brunt of responsibility should the product succeed or fail in the marketplace. All of these forces often initially land on the designer's doorstep, and so the designer often takes up the position of the visionary, the person tasked to reappraise the validity of the project and steer it into a more convincing and relevant direction, despite all the risk. They must navigate the shifting forces of the needs and desires of the company, all with the purpose of creating equilibrium: too expensive a product and it will not sell; too cheap in quality and expectations are not met; too loud a design and it is rebuffed; too neutral and desire is not created.

> 'Each building has to be beautiful, but cheap and fast, but it lasts forever. That is already an incredible battery of seemingly contradictory demands. So yes, I'm definitely perhaps a contradictory person, but I operate in very contradictory times.'
> Rem Koolhaas

There is a Japanese concept of beauty, called *wabi-sabi* (often translated as 'flawed beauty'), where things that are made should be imperfect and remain incomplete, in order to demonstrate impermanence. This kind of beauty is about leaving something open in the form of the object itself – a kind of unfinished circle for the human to complete. Bernard Leach, a twentieth-century, English studio potter, was a pioneer of this concept, as was Sōetsu Yanagi, a Japanese potter who practised a similar appreciation for 'objects born, not made'. Handmade objects are deliberately left open and rough, or unfinished. It could be the glaze that did not quite cover the whole pot, or a small deformation while it was fired.

These imperfections were seen not as a sign of artistic failure, but as a chance to further the concept that beauty is an open process. The more open an object is to use, the more beautiful it often feels. Sometimes, allowing nature to take its course can lead to timelessness.

This is very difficult to apply technically to mass production when industrial design relies upon consistency and repetition by its very nature. But there is value in the thought that beauty must remain open in its ideas rather than necessarily in its execution. The simplicity of an object, a piece of furniture or a system should appear open, so that the person occupying it will become part of it; part of its special and particular atmosphere.

Catalogue of Works

pp. *24–5*, 117
Door Tool 2001
Industrial Facility, England
Cast and machined aluminium
Prototype

fig. 02
Sofa 2001
Ryohin Keikaku (Muji), Japan
Plywood, foam, fabric
Production 2002–6

By 2001, Muji had already become known for simple designs using good materials: and for selling products at reasonable prices. We suggested continuing this formula for furniture, using plywood for its inherent strength and material honesty; and developing custom metal joints to avoid bolts piercing the wood surfaces, while still retaining a flat-pack shipping capability.

fig. 03
Bed 2001
Ryohin Keikaku (Muji), Japan
Plywood, tea board, fabric, steel
Production 2002–6

Following the sofa design, we applied our thinking to bed frames, introducing subtle bends and radii that are natural to plywood. The range proved popular enough to include both single and double beds.

pp. 147, *200–2*
Second Phone 2002
Ryohin Keikaku (Muji), Japan
ABS plastic, polypropylene
Production 2002–5
Good Design Award (G Mark), Japan, 2004; FX Award, England, 2004; D&AD, England, 2004; iF Gold Award, Germany, 2005
Permanent Collections: Art Institute of Chicago, USA; Israel Museum, Jerusalem; San Francisco Museum of Modern Art, USA

Magazine Stand 2002
Zing magazine, USA
Pinewood, hardboard
Limited edition of 30

Devon Dikeou, owner of art magazine *Zing*, asked us to design a display stand for shops and galleries. Two picture frames – one white, the other pinewood – are fixed together along one edge at 75 degrees. Each frame is able to display a magazine like a painting, with copies behind.

pp. *67*, 129
Landscape Camera 2002
Industrial Facility, England
Disposable camera mechanism, ABS and polycarbonate plastic
Prototype

pp. *76*, 131
Chantry Modern 2003
Knife sharpener
Taylor's Eye Witness, England
Plated, powder-coated and injection-moulded zinc; steel-carbide movement
Production 2004–
Design Plus Award, Germany, 2005
Permanent Collections: Die Neue Sammlung, Germany; Museum Angewandte Kunst, Germany; Art Institute of Chicago, USA

fig. 01
Hairglasses 2003
Hairband
Takumi, Japan
Nylon plastic
Production 2003–4

A hairband in the shape of a pair of sunglasses, sold only in Japan.

fig. 04
The Artist and Engineer 2003
Exhibition design
Maison et Objet, Paris, France
Collaboration with Philippe Boisselier

Under the title, *The Artist and Engineer*, this small exhibition put together two radically different approaches of the designers Ron Arad and Sam Hecht, placing Ron's supple forms alongside Sam's utilitarian objects. The exhibition was at the behest of Chantal Hamaide of *Intramuros* magazine.

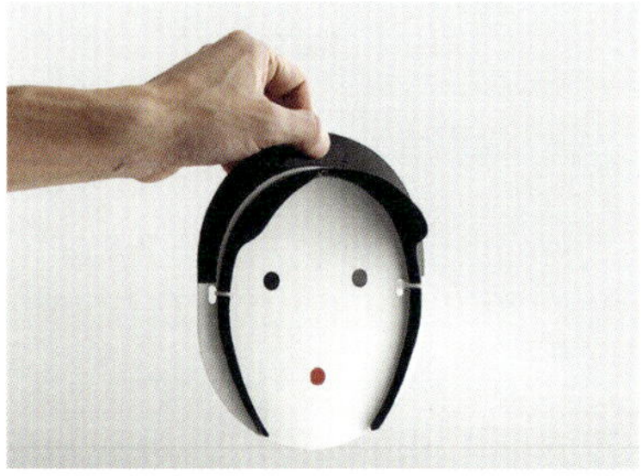

01

02

03

04

05

06

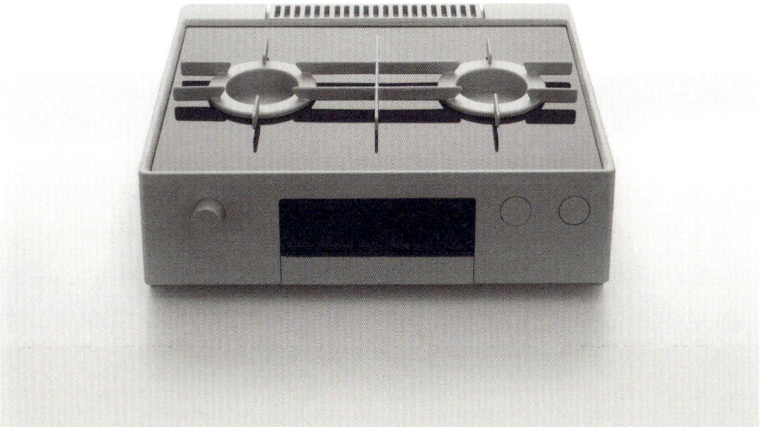

07

08

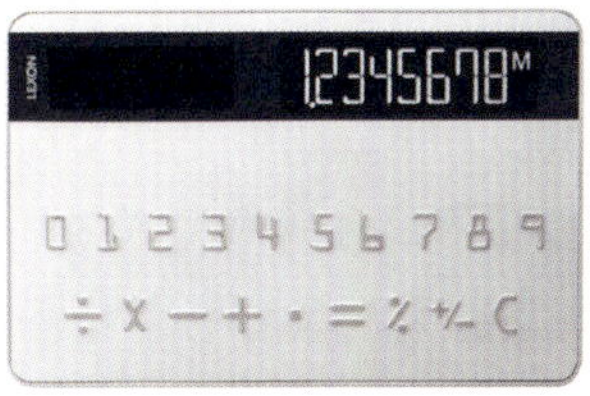

09

10

fig. 09
Credit 2003
Pocket calculator
Lexon, France
ABS plastic injection moulding, embossed polyester
Production 2004–6
Permanent Collection: Centre Pompidou, France

René Adda, director of the French company, Lexon, is a businessman who adores design. He visited us with a proposal to design personal accessories and we decided to use it as an opportunity to collide the familiar with the electronic. In this example, we were able to replicate the OCR font used on credit cards as the interface for a solar calculator.

fig. 10
Give and Take 2003
Business card holder
Lexon, France
Anodized extruded aluminium
Production 2004–6
Permanent Collection: Centre Pompidou, France

A two-sided business card holder made from thin-wall aluminium extrusion, with one side for receiving and the other for giving.

fig. 08
Mathmos 2003
Store design
Mathmos, England
Cardboard tubing, plywood

The lighting company, Mathmos, was around the corner from Industrial Facility's first office in Golden Lane, London. They asked us to design their showroom and the solution was to install a series of custom-manufactured cardboard tubes of different diameters and lengths, corresponding to the light levels that each product emits. To one side was an enormous 3 m (9 ft 8 in) tube that provided information about the company.

pp. 142, *178–9*
Flexlamp 2003
Pendant lamp
Droog, the Netherlands
Moulded silicone
Production 2004–13
Permanent Collection: Israel Museum, Jerusalem

fig. 05
Low Sofa 2003
Ryohin Keikaku (Muji), Japan
Plywood, foam, fabric
Production 2004–6
Permanent Collection: Israel Museum, Jerusalem

Muji noticed the growing popularity of 'low' sofas – a type of sofa that is close to the floor. Ours connects the back frame to an armrest that can be rotated independently of the seat. The advantage of this is flexibility in use and arrangement so that seats can be put together with other modules, and users can decide whether there is an armrest or not.

fig. 11
Equipment 2003
Compact kitchen
Whirlpool, Italy
Anodized aluminium, Corian, glass
Prototype
Red Dot Award, Germany, 2005

Equipment coincided with a philosophy in the studio that we termed 'voluntary simplicity': a concern to break the notion that luxury meant large, more and expensive. Instead, we thought of people volunteering to have less, but made well. We took advantage of a request from Whirlpool by designing a kitchen that was 2.3 × 1.6 m (90.5 × 63 in), as though it were a single appliance. Flexible and multifunctional, the design involved the same manufacturing methods used in car production.

fig. 07
Tabletop Cooker 2003
Rinnai, Japan
Powder-coated steel, induction glass
Prototype

The German company, Schott, which manufactures glass for electric stovetops, asked if we would work with Nagoya-based Rinnai to help modernize a common product in Japan, called a 'Gas Table'. These products are plagued by dirt and our idea was to clean it all up. In the process, we created a new feature whereby a pan could slide from one burner to a 'neutral' zone in the middle.

fig. 12
Chair Stool 2003
Guest chair with integrated stool
Strafor (Steelcase), France
Glass-reinforced plastic
Prototype

Strafor commissioned us to examine what a 'guest' chair could be for the modern office. We thought that rather than a stackable chair, it could be a chair that holds its own stool. If more than one guest arrives, you could release the stool from underneath. The design remained as a prototype.

pp. *77*, 131
IF4000 2004
Knives
Taylor's Eye Witness, England
High carbon stainless steel, melamine and polyester handles
Production 2005–15
iF Gold Award, Germany, 2006; Design Plus Award, Germany, 2006; Grand Designs Award, England, 2008
Permanent Collections: Museum Angewandte Kunst, Germany; Art Institute of Chicago, USA

pp. *34–5*, 119
City in a Bag 2004
London, Paris, Tokyo, Kyoto, New York, Italy, Germany, China, Taiwan, Barcelona, Berlin
Ryohin Keikaku (Muji), Japan and Europe
FSC sustainable and renewable beechwood
Production 2004–15
Permanent Collection: Art Institute of Chicago, USA

fig. 06
Foot Care Tools 2004
Ryohin Keikaku (Muji), Europe
Cedarwood
Production 2004–6

pp. *91*, 135
Paper Cup 2004
Takeo, Japan
Single-wall, waxed, Japanese tracing paper
Commission for *Haptic* exhibition, curated by Kenya Hara

11

12

13

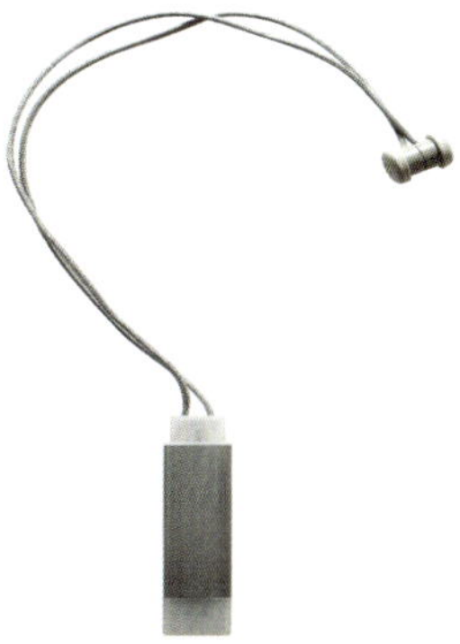

14

15

16

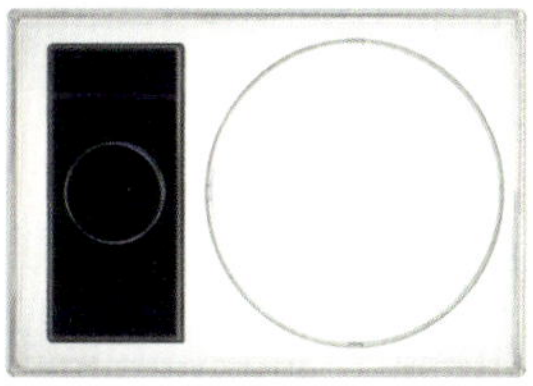

17

fig. 15
Duo 2004
Ballpoint and highlighter pen
Lexon, France
Extruded aluminium, silicone
Production 2004–5
Permanent Collection: Centre Pompidou, France

Two pens can be kept together using a double cap with an integrated clip.

fig. 13
Key 2004
Torch
Lexon, France
ABS plastic injection moulding
Production 2004–8
Permanent Collection: Centre Pompidou, France

Similar to the Credit solar calculator that references a credit card [p. 258], Key is a small LED torch with the outline of a key.

fig. 16
Mouse 2004
Lexon, France
ABS plastic injection moulding
Production 2004–7
Permanent Collection: Centre Pompidou, France

We dramatically simplified the appearance of an optical mouse by turning the plastic housing into the buttons themselves, by way of a slit in the middle. This allowed the housing to act similarly to two hinges: a method used by countless competitors soon after the product launched.

fig. 14
Soundgum 2004
MP3 player
Lexon, France
Extruded aluminium
Production 2004–5
Permanent Collection: Centre Pompidou, France

For Lexon's MP3 player, we created several small surprises: the first was to make it essentially like a USB stick; the second was to magnetize the headphones so that they could clip together, allowing the device to be worn around the neck when not being used. Like the optical mouse before it, this small interjection of magnets in the headphones became a common feature.

fig. 18
Once 2004
Watch
Lexon, France
Soft ABS plastic injection moulding
Production 2004–6
Permanent Collection: Centre Pompidou, France

A digital watch where the display itself is one big physical button used to access different mode operations.

fig. 17
Spot Light 2004
Magnifying glass with light
Lexon, France
Soft ABS plastic injection moulding
Production 2004–5
Permanent Collection: Centre Pompidou, France

A small magnifying glass with an LED light source. The two functions were kept visually separate, but enclosed within the same acrylic block.

fig. 19
Twice 2004
Alarm clock
Lexon, France
ABS plastic and acrylic injection moulding
Production 2004–5
Permanent Collection: Centre Pompidou, France

Similar to the Once watch, the Twice alarm clock's display is one large button that both operates the light and stops the alarm from sounding.

fig. 20
Status 2004
Wall clock with outdoor weather sensor
Lexon, France
Soft ABS plastic injection moulding
Production 2004–6
Permanent Collection: Centre Pompidou, France

A wall clock that comes with an external battery-powered weather sensor (the sensor neatly clips into the back of the clock when not being used). To allow the clock to be installed at any height, it is able to rotate up and down within its housing.

18

19

20

21

22

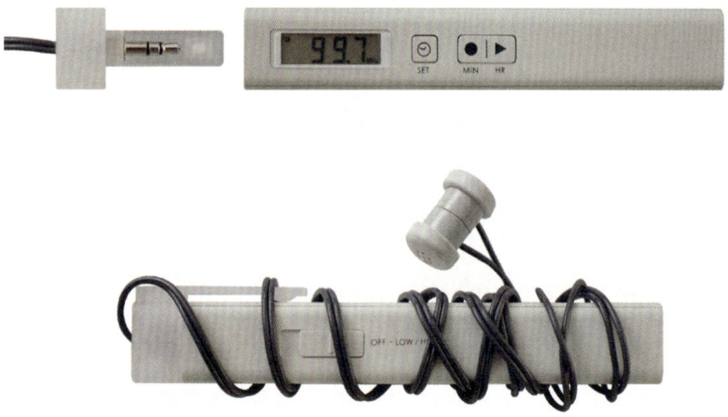

23

fig. 23
Radius 2004
Radio and headphones
Lexon, France
Soft ABS plastic injection moulding
Production 2004–6
Permanent Collection: Centre Pompidou, France

A small personal radio where the headphone plug also acts as a shirt clip. The headphones use the same magnetic attachment method that we used in the Soundgum MP3 player.

fig. 21
Playing Cards 2004
Ryohin Keikaku (Muji), Japan and Europe
Recycled card
Production 2005–8

Muji's playing cards are an exercise in just how much imagery can be removed from the cards, while still allowing enough information for games to be played.

fig. 22
Soft Light 2004
Industrial Facility, England
Cast silicone rubber
Limited edition of 10

pp. *72–3*, 129
Coffee Maker 2005
Ryohin Keikaku (Muji), Japan
Polypropylene and steel plate
Production 2006–12
iF Gold Award, Germany, 2007

pp. *74–5*, 130
IF4000 2005
Knife rack
Taylor's Eye Witness, England
Laminated bamboo, beech and oak
Production 2006–12, 2014–
Design Plus Award, Germany, 2007; British Housewares Gold, England, 2007
Permanent Collections: Die Neue Sammlung, Germany; Museum Angewandte Kunst, Germany

pp. *46–7*, *52*, 121
Picturemate 2005
Colour photo printer
Seiko Epson Corporation, Japan
ABS plastic, polyester-reinforced handle
Production 2006–
iF Gold Award, Germany, 2006; Good Design Award (G Mark), Japan, 2006
Permanent Collection: Israel Museum, Jerusalem
Collaboration with Epson Design

pp. *54–6*, 124
Fan 2005
Ryohin Keikaku (Muji), Japan
Polypropylene and ABS plastic, steel wire
Production 2006–10
iF Award, Germany, 2007

fig. 24
Shopping Bag 2005
Ryohin Keikaku (Muji), Europe
Coloured Tyvek cloth
Production 2005–6

A simple shopping bag made of the paper-based material, Tyvek. The handle is made from a loop of fabric that provides both a short version for holding it by hand, and a shoulder version that can be pulled out from the inside.

fig. 25
Shoe Kit 2005
Ryohin Keikaku (Muji), Europe
Leather, nylon fabric
Production 2005–6

A small, zipped case in the shape of a shoe insole, which contains a polishing kit.

fig. 30–1
Tea Candle 2005
Ryohin Keikaku (Muji), Japan and Europe
Borosilicate glass
Production 2005–10

Applying the concept of utility to the tea light resulted in this small, glass storage vessel that can hold up to five candles. The lid is lifted up and rotated to hold the candle. It was originally produced for the European market in 2005, followed by a Japanese version in 2009, made from frosted glass.

fig. 26
Surround Sound Eyewear 2005
Royal National Institute for the Deaf
Acrylic plastic
Commission from the Victoria and Albert Museum, England, 2005

We thought it conceivable that someone who is hard of hearing but who has perfect eyesight may prefer to wear glasses as a hearing aid, even though the lenses are of no benefit. That was the start of a concept that led us to a technology called super-directive beam forming, developed by Professor Marinus Boone of Delft University. By positioning a set of four microphones on either side of the glasses' arms, hearing capability increases dramatically.

fig. 29
Glassware 2005
Ryohin Keikaku (Muji), Japan
Lead-free crystal
Production 2006–8

Everyday glasses for water, wine, soft drinks and Prosecco. The defining elements were a very thin edge, made possible by the factory's knowledge of crystal, and a single, stackable form that was applied equally to all of the glassware.

fig. 27–8
Hooks and Towel Hangers 2005
Ryohin Keikaku (Muji), Japan
Soft ABS plastic, magnets
Production 2006–9
Good Design Award (G Mark), Japan, 2006

A collection of hooks that use suction or magnets as a method of connection, each made from one seamless moulding.

24

25

27

26

28

29

30

31

32

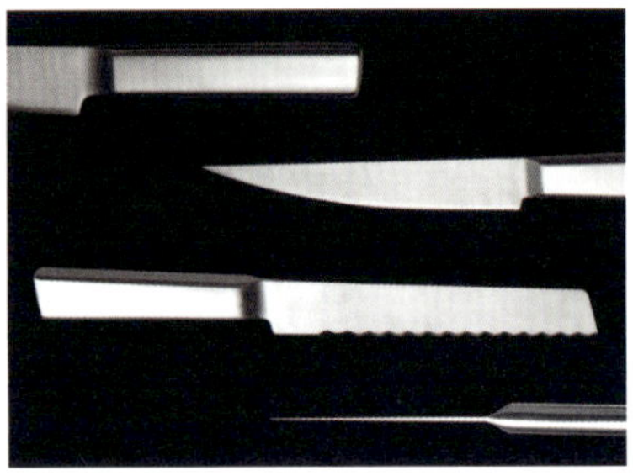

33

34

35

fig. 32
Found Made Thought 2005
Exhibition design
Permanent Collection: Israel Museum, Jerusalem

In 2005, we were still sharing an office with the designers Dunne and Raby, and we both simultaneously produced exhibitions in two adjoining sections of the same gallery space. *Found Made Thought: The Work of Industrial Facility* showed design that had been created within the constraints of commercial considerations; while *PopNoir: Critical Designs Curated by Dunne and Raby* displayed design that was speculative or conceptual in approach.

fig. 33
IS3000 2005
Kitchen knives
Taylor's Eye Witness, England
High carbon stainless steel
Production 2006–8

The design direction for IS3000 was taken from how a metalsmith uses a mechanical hammer to forge steel bars to make them thinner. The forms of the knives replicate the feeling that a single stainless steel bar has been pounded at one end to form the blade, with the other left untouched to act as the handle. In reality, the blade is welded to a hollow handle, but the result appears as one single, seamless and hygienic material.

fig. 34
Light Stool 2005
Ryohin Keikaku (Muji), Japan
Expanded polystyrene EPP, foam and fabric
Production 2005–7

Expanded polystyrene was gradually entering the furniture market as a way of creating large, lightweight structures. In this example, we used it to make a light footstool.

fig. 35
Spray watering can 2005
Ryohin Keikaku (Muji), Japan
Blow-moulded polypropylene
Prototype

After Sam found an odd watering can in Asia that integrated a spray function, he showed it to Muji as a lighthearted joke – only to then be asked if he could make it work. To do this, the design was dramatically reduced in scale for use on small balconies. It remained a concept.

pp. *59–61*, 126
Passport Memo 2006
Ryohin Keikaku (Muji), Japan
Coloured card, staple-bound pages
Production 2006–
Permanent Collection: Art Institute of Chicago, USA

pp. *102*, 137
Jetlag 2006
Travel clock
Takumi, Japan
ABS plastic injection moulding
Production 2007–12
Permanent Collection: Israel Museum, Jerusalem

pp. *106–9*, 138
Cordless Phone 2006
Industrial Facility, England
Polished ABS plastic
Prototype

pp. 145, *196–7*
Little Disk 2006
Hard drives
LaCie, France
High-pressure ABS plastic
Production 2007–12
Creative Award, Japan, 2007; iF Award, Germany, 2008; Red Dot Award, Germany, 2008; Homes and Gardens Technology Award, England, 2008
Permanent Collections: Centre Pompidou, France; Israel Museum, Jerusalem

pp. 148, *203–5*
Projector 2006
Seiko Epson Corporation, Japan
ABS plastic, acrylic, glass-filled nylon
Prototype

fig. 43
Football 2006
Design Museum, England
Football, phosphorescent dye
Commission for *Football Fever*,
Design Museum, England, 2006

To celebrate the 2006 World Cup, the Design Museum in London made an exhibition called *Football Fever*. Our contribution was to apply a luminous, 'glow-in-the-dark' coating to the ball. With a camera flash, it glowed for a couple of minutes at a time.

fig. 38
Step 2006
Induction and gas stovetop
KitchenAid, Italy
Induction glass, stainless steel,
iron casting
Production 2007–11
iF Gold Award, Germany, 2008;
Red Dot Award, Germany, 2008

We suggested to KitchenAid that they bring together two cooking technologies – gas and induction – into a single stovetop. Each provides varying degrees of control, and we articulated them into a new appliance topography, in which gas (for heavy cooking) is higher and to the rear, and induction (for fine cooking) is kept at the front, flush with the work surface.

fig. 45
Suburbia in a Bag 2006
Ryohin Keikaku (Muji), Europe
FSC sustainable and renewable
beechwood
Production 2006–8
Permanent Collection: Art Institute
of Chicago, USA

fig. 40
Pen Magnet 2006
Ryohin Keikaku (Muji), Japan
Polypropylene plastic moulding
with inserted magnet
Prototype

A simple plastic rod that shares the same dimensions as a pen: instead of holding ink, it holds a magnet for attaching paper clips to its tip.

36

39

40

37

41

42

38

fig. 46
Fractions in a Bag 2006
Ryohin Keikaku (Muji), Japan
Beechwood, silk-screening
Production 2007–8

In 2006 we started to create a series of educational toys inspired by the Montessori method of teaching. Simple blocks of wood are used as a method for explaining volume, percentages and fractions.

fig. 47
Percentages in a Bag 2006
Ryohin Keikaku (Muji), Japan
Beechwood, silk-screening
Production 2007–8

fig. 44
Volume in a Box 2006
Ryohin Keikaku (Muji), Japan
Beechwood, silk-screening
Production 2009–10

fig. 41
Chalk Car and Truck 2006
Ryohin Keikaku (Muji), Japan
Beechwood, chalk paint
Production 2007–8

Injection Pens 2006
Novo Nordisk, Denmark
Polypropylene plastic
Prototypes

Over many years, the studio has enjoyed engaging with the medical company Novo Nordisk to instill simplified design awareness across all parts of the company.

fig. 36
A-Size Ruler 2006
Ryohin Keikaku (Muji), Europe
Silk-screened beechwood
Prototype

A measuring ruler that did not reach production but that was nevertheless a favourite in our office. It calls out the DIN sizes of A4, A5 and A6.

fig. 42
Design Language 2006
Orange / France Telecom, France
ABS plastic, silicone
Prototypes

pp. *53*, 123
Circumference 2007
Watch
Ryohin Keikaku (Muji), Japan
Stainless steel, ionized plating, crystal glass
Production 2009–11
iF Award, Germany, 2012
Permanent Collection: Art Institute of Chicago, USA

pp. *79*, 133
Wine Opener 2007
Ryohin Keikaku (Muji), Japan
ABS plastic, ionized stainless steel
Production 2008–13
Good Design Award (G-Mark), Japan, 2009
Permanent Collection: Art Institute of Chicago, USA

pp. 142, *177*
Lite Plug 2007
Emergency torch, night light
Takumi, Japan
ABS plastic with fireproof coating, polycarbonate
Production 2008–13
Permanent Collection: Art Institute of Chicago, USA

pp. 144, *190–1*
Toilet Brush and Bin 2007
Ryohin Keikaku (Muji), Japan
ABS plastic, polypropylene
Production 2007–
Good Design Award (G Mark), Japan, 2008

pp. 151, *216–17*
Two Timer 2007
Wall clock
Established & Sons, England
Tempered glass, aluminium, quartz movements
Production 2008–12
Creative Review Best Graphic Product, England, 2009; Grand Designs Award, England, 2010
Permanent Collections: San Francisco Museum of Modern Art, USA; Israel Museum, Jerusalem

pp. 144, *192*
Bath Radio 2007
Ryohin Keikaku (Muji), Japan
ABS plastic injection moulding
Production 2008–14

pp. 154, *232–3*
Shredder 2007
Industrial Facility, England
ABS plastic injection moulding
Prototype

Photo Viewer 2007
Seiko Epson Corporation, Japan
ABS plastic, tempered glass
Prototype

The design of a professional mobile tablet for the display of SLR photographs.

fig. 39
Beam 2007
Pendant and floor lamp
Established & Sons, England
Blown glass, spun steel
Production 2007–12
Grand Designs Award, England, 2008

When a child draws a typical lampshade, they sometimes continue to draw the light emitting from a lamp too – as if light were an object. This idea was interpreted into Beam, where a spun-metal shade sits on top of a larger cone of translucent blown glass.

fig. 37
Encord 2007
Home office table and storage
Herman Miller, USA
Cast aluminium, polyester-coated MDF
Production 2008–12

A table that combines the needs of home life with work. It has two tabletops, one beneath the other. The gap between them can be used to store laptops, papers, or the mass of cords and cabling that never seems to go away. The lower surface also extends beyond the top by 30.5 cm (12 in), acting as a side table for peripherals.

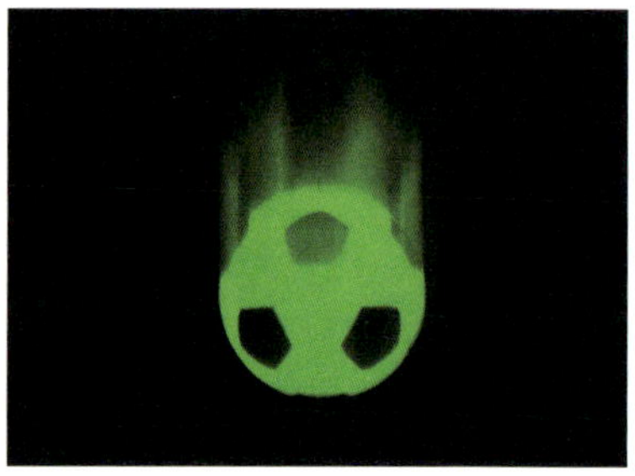
43

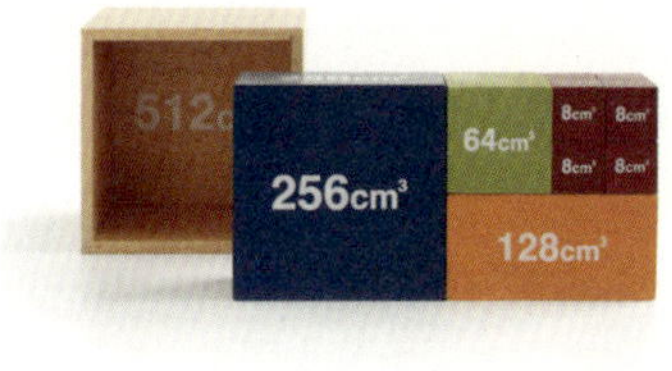

44

45

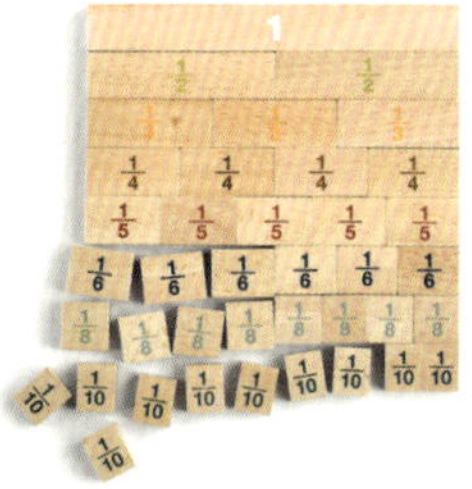
46

47

48

49

fig. 50
Lo Cost Cutlery 2007
Taylor's Eye Witness, England and Ryohin Keikaku (Muji), Europe
Stainless steel
Production 2007–9
Permanent Collections: Art Institute of Chicago, USA; Design Museum, England

In talks with director, Alastair Fisher, we pondered the subject of low-cost cutlery sold in supermarkets and remarked how it tends to mimic expensive silverware. We decided to reference the opposite: to elevate the vernacular of disposable plastic cutlery instead. In our workshop, we cast metal into moulds made from plastic cutlery that we had gathered, and once made in metal it showed us just how beautiful these designs were. We then went through a process of refining and editing them into a four-piece place setting, including a spork (a combination of spoon and fork).

fig. 48
Space in a Bag 2007
Ryohin Keikaku (Muji), Japan
FSC sustainable and renewable beechwood
Production 2007–9
Permanent Collection: Art Institute of Chicago, USA

fig. 49
Cookie Cutter 2007
Ryohin Keikaku (Muji), Europe
Steel plate
Production 2007–9

fig. 51
Multi-Function Printer 2007
Seiko Epson Corporation, Japan
ABS plastic
Prototype

A study was made with Epson into how best to approach a printer that can also make scans and photocopies. The solution was to divide the functions visually into horizontal bands that instantly made the product seem more understandable.

fig. 52
City Clock 2007
Ryohin Keikaku (Muji), Japan
Painted beechwood, silk-screened paper
Production 2008–9

fig. 55
Pixi 2007
Social camera
Matsushita Electric Industrial (Panasonic), Japan
ABS plastic injection moulding
Prototype
Collaboration with Madori Kuroda and Durrell Bishop

Pixi was a conceptual project that allowed a mobile phone to slip into a bright orange camera case. You could take a picture and send it to a social network by way of a wireless interface on the back. It predated the rise of Instagram. The name Pixi came from Mixi, the largest social network in Japan at the time.

fig. 54
Some Recent Projects 2008
Exhibition concept and design
Design Museum, England
Collaboration with Graphic Thought Facility

Instigated by Deyan Sudjic and curated by Gemma Curtin, this was Industrial Facility's first solo exhibition in England. In one section, called 'Some Recent Projects', it displayed a selection of the studio's designs, exploring how its work tries to push the boundaries of materials and methods. In another, called 'Under a Fiver', it presented almost 100 collected objects from around the world that have influenced the studio's work. A large, glass-covered bookshelf separated the two sections, doubling as both a practical means for storing and dispensing the gallery guide, as well as for displaying the exhibition signage.

pp. *92–5*, 136
Table, Bench, Chair 2008
Benches
Established & Sons, England
English oak with stained ash
Production 2009–14
Bo Bedre Award, Denmark, 2010; Wallpaper Design Award, England, 2010

pp. 140, *160–3*
Bell 2008
Alarm clock
Takumi, Japan
Pressed steel, ABS plastic
Production 2009–14
Permanent Collection: San Francisco Museum of Modern Art, USA

50

51

52

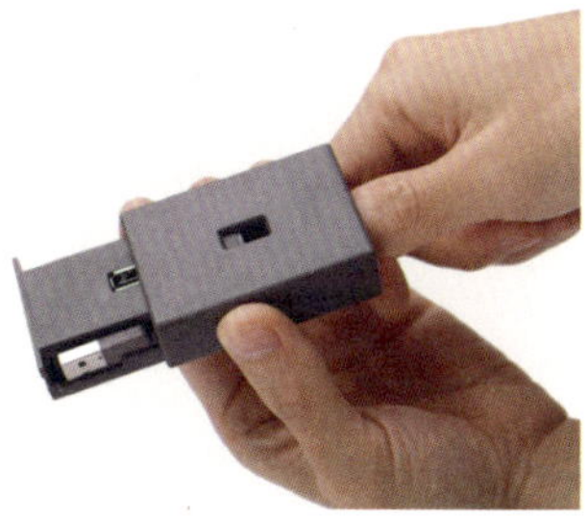

53

54

fig. 56
Core 7 2008
USB hubs
LaCie, France
ABS plastic
Production 2009–12
Red Dot Award, Germany, 2009

The Core 7 and Core 4 USB hubs represent the second project we undertook for LaCie. The Core 7 hub can be used on the desk or wall mounted, for plugging in USB devices. At the rear of the hub are two compartments that house a regular and a mini USB cable, along with six USB sockets. The seventh is on the top surface. The Core 4 hub design is similar in approach but is a mobile device, providing four USB sockets.

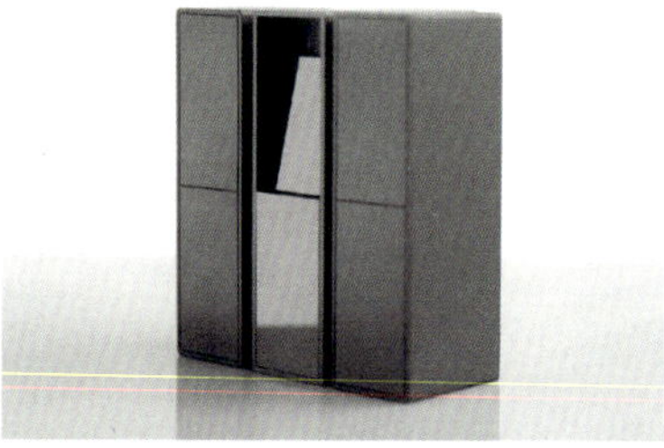
57

fig. 53
Core 4 2008
Mobile USB hub
LaCie, France
ABS plastic
Production 2009–12
Red Dot Award, Germany, 2009

55

pp. 133
Wine Peg 2008
Ryohin Keikaku (Muji), Japan
ABS plastic
Production 2008–11
Good Design Award (G-Mark), Japan, 2009
Permanent Collection: Art Institute of Chicago, USA

58

fig. 58
Mizu 2008
Washbasin collection
Porcelain
Villeroy & Boch, Germany
Prototype

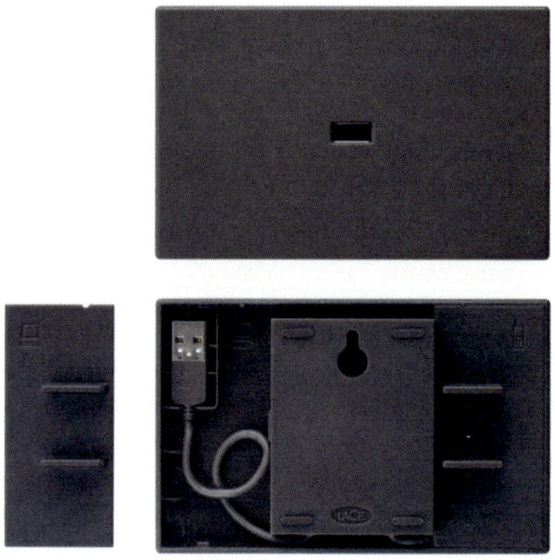
56

fig. 57
Landscape of HiFi 2008
Design Strategy
Matsushita Electric Industrial (Panasonic), Japan
Prototype
Collaboration with Yohei Yamamoto, Panasonic Design

Television 2009
LG Electronics
ABS plastic, pressed aluminium
Prototype

An investigation into television design, where the TV could now be considered to be no more than a large media monitor.

fig. 59–60
Parfum Concession 2009
Kiosk and display accessories
Issey Miyake Parfums, France
Acrylic, lacquered MDF

The Issey Miyake kiosk design was both architecture and product display. We wanted to express the idea that while the sensory experience of perfume is accessible, the experience can also be highly curated. Box displays, perfume stands, test placements and trays were combined to create a fresh and informal presentation on top of forms that related to packaging itself. The design was first launched at Galeries Lafayette in 2009, and then rolled out globally.

fig. 62
K9 Post Office Kiosk 2009
Design Museum, England
Solar glass, iroko planking
Commission for *Super Contemporary*, Design Museum, England

As part of the exhibition *Super Contemporary*, we carried out extensive research into the feasibility of a Post Office that shares the same dimensions as a telephone box. The kiosk provides a range of services including selling stamps and TV licenses, and making passport applications. The project was not about replacing the Post Office but about supporting a network that is under great economic pressure. On the screen inside is a 'real' member of staff displayed at 1:1 scale, who can engage with the customer using real-time, centralized video conferencing.

fig. 69
Dip 2009
Bowl and vase
Established & Sons, England
Black glazed ceramic
Production 2009–10

fig. 67
Compact Barbecue 2009
Ryohin Keikaku (Muji), Japan
Steel pressing with zinc coating
Prototype

pp. *26–7*, 117
Twinstopper 2009
Doorstop
Droog, the Netherlands
Injection-moulded silicone
Production 2010–

fig. 65
Jersey Dishes 2009
Jersey Pottery, Jersey
Glazed porcelain, stencil printing
Production 2007–11
Best Kitchen Product Award, *Elle Decoration*, 2008

fig. 61
E-Motion 2009
Mobile phone
Matsushita Electric Industrial (Panasonic), Japan
Injection-moulded plastic
Prototype

A mobile phone design project instigated by Panasonic for the Japanese service provider NTT DoCoMo.

fig. 64
Bubble 2009
Laptop bags
LaCie, France
Neoprene fabric
Production 2010–13

Bubble bags for laptops, hard drives and tablets that are made from neoprene, with a texture that mimics exactly the dimensions of plastic bubble wrap.

fig. 68
Compression 2009
Television
Hitachi, Japan
Stainless steel ABS plastic
Prototype

As part of the Product as Landscape manifesto that we had articulated in a book of the same name, we suggested to Hitachi that the physical aspect of a television could be reliant more on its surroundings, proposing that it could lean against a wall, thereby not requiring a stand. Hitachi was willing to go along with it. Prototypes were produced where the bezel that surrounds the screen continued to the floor in the form of two thin, stainless-steel legs that allowed it to either lean against or be attached flat to the wall. Power cables were able to run down the inside of each leg.

59

60

63

61

64

62

65

69

fig. 63
Cozy 3 2009
Laptop bag
LaCie, France
Neoprene fabric, olefin plastic
Prototype

Following the LaCie Bubble bags, we developed a neoprene, zipped pouch for laptops. Instead of integrating the handles into the neoprene, we gave it a semi-rigid exoskeleton that slipped over the pouch.

pp. *62–5*, 127
Branca 2010
Chair and table
Mattiazzi, Italy
CNC ash, oak with natural stains
Production 2011–
Designs of the Year Award, England, 2011; Habitare, Finland, 2011
Permanent Collections: Art Institute of Chicago, USA; San Francisco Museum of Modern Art, USA; Designmuseo, Finland; Design Museum, England; Victoria and Albert Museum, England

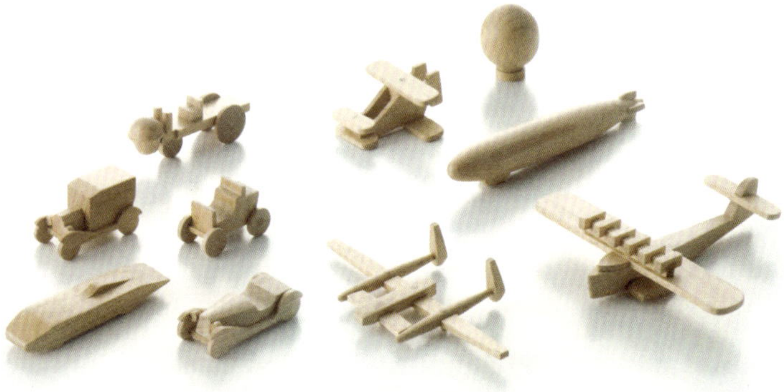

66

70

71

pp. *103–5*, 138
Timer Lamp 2010
Ryohin Keikaku (Muji), Japan
Soft silicone, ABS plastic
Production 2011–
iF Award, Germany, 2012
Permanent Collection: Art Institute of Chicago, USA

fig. 70
Phone Puzzle 2010
Ryohin Keikaku (Muji), Japan
Silk-screened beechwood
Production 2011–13

fig. 71
Apple Puzzle 2010
Ryohin Keikaku (Muji), Japan
Silk-screened beechwood
Production 2011–13

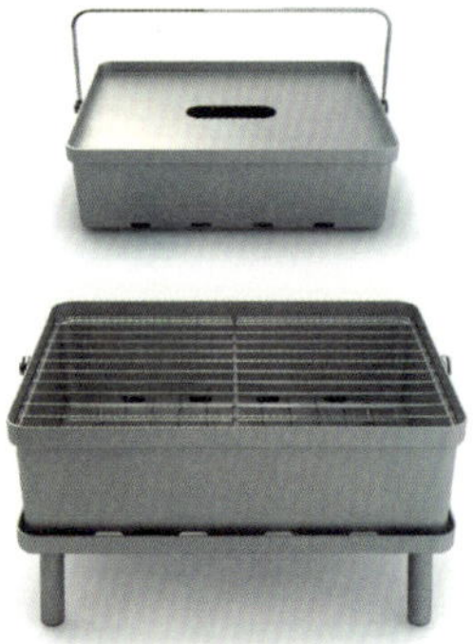

67

68

fig. 66
Transport in a Bag 2010
Ryohin Keikaku (Muji), Japan
Beechwood
Production 2011–13

Safari in a Bag 2010
Ryohin Keikaku (Muji), Japan
Beechwood
Production 2011–13

72

73

74

pp. *57*, 125
Cycle Shirt 2010
Margaret Howell, England
Pre-washed cotton
Limited edition of 200
Permanent Collection: Art Institute of Chicago, USA
Collaboration with Margaret Howell

pp. *58*, *60–1*, 125
Pentagon 2010
Fountain and ballpoint pen
Elephant & Coral, Singapore
Polished ebonite
Limited edition of 300
iF Gold Award, Germany, 2011
Permanent Collection:
Art Institute of Chicago, USA

Silver Ruler 2010
Ryohin Keikaku (Muji), Japan
ABS plastic
Production 2011–

In Japan, an ageing population has become disproportionately large due to people living longer. They call this the 'silver' generation and Muji were interested in the design of objects that were sensitive to this new reality. One was a ruler that we made black with a larger font size printed in white, dramatically improving legibility. Another was a diary, where each date fills an entire square that can be written over.

fig. 76
Silver Diary 2010
A5 and A6 diary
Ryohin Keikaku (Muji), Japan
Jeans label cover, printed paper
Production 2011–12

fig. 72
Information Watch 2010
Ryohin Keikaku (Muji), Japan
Polycarbonate plastic and silicone
Prototype

After Muji had produced the Circumference watch [p. *53*], they asked us to investigate what a digital watch could be. It was in the summer of 2010 and Rafael Nadal was in the finals of Wimbledon. The scoreboard was a complete panel of information in pixelated form, and we thought this a reasonable reference for a watch. Time, seconds, date and alarm signal are all presented together on one screen, Wimbledon-style.

fig. 80
Omnipresence 2010–15
Technology investigations: Car, Travel, Retail, Communication, Office, Home, Mobile
Research in Motion (BlackBerry Limited), Canada
Prototypes

Over a period of five years, design director Todd Wood commissioned us to produce conceptual designs, along with writings, each time investigating a different topic that was being confronted with change because of mobile technologies. The projects were intellectually enjoyable, partly because of Todd's ability to protect the projects from too much corporate interference. The result was that, twice a year, executives could immerse themselves into new potentials, which were opinionated and thought-provoking, resulting in a number of patents.

fig. 74
Dryerhair 2010
Hairdryer
Industrial Facility, England
Polycarbonate plastic
Commission for Wallpaper
Handmade, Italy, 2010

The design of hairdryers has remained curiously static over the years, apart from ever more arbitrary styling. To redress this, *Wallpaper* asked us to rethink the ubiquitous blow-drying tool. Looking at how the object fits into a room or cupboard, we looked at the storing of the cable – a particular annoyance. The design incorporates a translucent cavity into which the cable can neatly coil when not in use, with the electrical plug also acting as a lid. The switch, meanwhile, was tucked out of sight in the base of the handle.

fig. 79
Air Freshener 2010
Ryohin Keikaku (Muji), Japan
ABS plastic
Prototype

Future Craft 2010
Design language
Matsushita Electric Industrial (Panasonic), Japan

A project that created a kind of design manual for how to locate and then design what is essential for any domestic electrical product.

fig. 75
Cantilever 2010
Electronic piano
Yamaha, Japan
Corian
Prototype

Cantilever is an electronic piano that uses the synthetic material Corian to counter any perception of inferiority in digital sound output. Its two simple volumes can only be achieved with electronic components, creating a form that brings the electronic piano into modernity, stripping away all classical references. The top surface gently slides back to reveal the piano keys, with the eighty-ninth key serving as a removable remote control. Shortly after its public release, Yamaha retreated from its development due to economic conditions.

fig. 73
An Alarm 2010
Alarm watch
Takumi, Japan
Stainless steel, crystal glass
Production 2010–13
Permanent Collection:
San Francisco Museum of Modern Art, USA

An analogue alarm watch, which adopts the interface of an alarm clock. The alarm is set simply by rotating the winder to the desired time and lifting the alert button, which is found in the same position as a typical alarm clock's. When the alarm is activated, a small motor creates a vibration on the wrist, similar to a mobile phone.

pp. *28–33*, 118
Wireframe 2011
Sofa
Herman Miller, USA
Coated steel wire, injection-moulded webbing, multiple-density foams, fabric
Production 2013–
Design Guild Award, England, 2014

fig. 78
Solar Storage 2011
Battery charger
Takumi, Japan
ABS plastic injection moulding, glass
Prototype

A small glass vessel with a lid that acts as a solar charger. As batteries are charged, they can be stored in the glass jar.

pp. 151, *218–21*, *226–7*, *230–1*
Locale 2011
Office furniture system
Herman Miller, USA
Cast aluminium, recycled felt, steel, Formcoat
Production 2014–
Best of NeoCon, USA, 2013; HiP Award, USA, 2014; Metropolis Award, USA, 2014; IDEA Gold Award, USA, 2014; Büro Best of the Best, Germany, 2015; Good Design Award, Australia, 2015; iF Gold Award, Germany, 2016

fig. 81
Solo, Vol. 21 2011
Artist's folio case
Toluca Editions, France
Corian, tempered glass
Limited edition of 20
Collaboration with Jorge Macchi and Alan Pauls

Each year, Toluca Editions commissions a folio that is a collaboration between an artist, a writer and a designer. The artist prepares six photographs, the writer contributes poetry, and the designer proposes a box to contain it. Solo, Vol. 21 holds photographs by Argentinian artist, Jorge Macchi, which feature urban trees that have grown and swelled to subsume manufactured elements like metal signage. Unlike previous editions that only boxed the photographs, we wanted to be able to display them as well. A Corian and glass case in the shape of a parallelogram can be hung (gravity ensures that it remains closed), while also storing the remaining five photographs. It was expertly crafted by Hubert Weinzierl.

fig. 77
Re-Mouse 2011
Remote control / optical mouse
LaCie, France
ABS plastic
Prototype

Similar in form to the Little Disk hard drives [pp. *196–7*], Re-Mouse involves a cap that houses a USB plug, which acts as a wireless transmitter. The mouse can also be used as a remote control for making screen presentations.

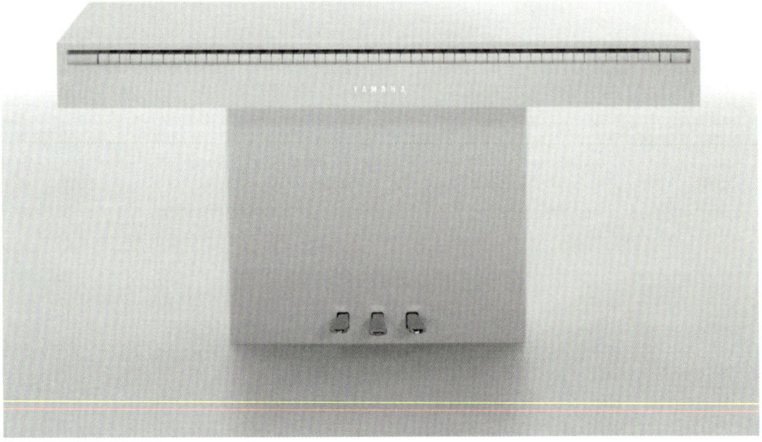
75

76

79

77

80

78

81

pp. *36–7*
Magnetic Tape 2011
Ryohin Keikaku (Muji), Japan
Silk-screened flexible magnet
Production 2012–13

A flexible magnet that borrows the appearance of torn tape, used to hold up notes.

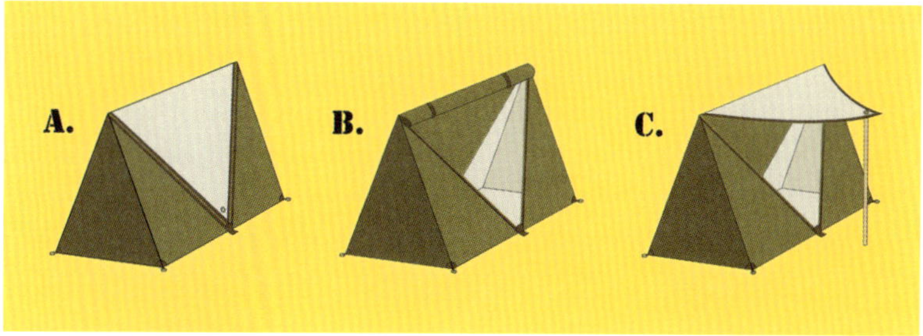

85

82

fig. 89
Human Puzzle 2011
Ryohin Keikaku (Muji), Japan
Silk-screened beechwood
Production 2011–13

fig. 86–7
Picnic Bag 2011
Ryohin Keikaku (Muji), Europe
Waterproof nylon with zips
Production 2011–12

A bag that can be used to carry picnic supplies, before then being unzipped to create a soft-padded picnic mat.

86

87

Layout 2011
Office furniture system
Herman Miller, England
Rolled steel, aluminium extrusion, laminate
Production 2014–

During the development of Locale [pp. *218–21*], Herman Miller, England, asked us to design a conservative furniture system that covered both desking and wall based shelving. This materialized into what is now called Layout.

83

pp. *88–90*, 135
Alumi 2012
Chair and table
Tectona, France
Extruded, pressed and powder-coated aluminium
Production 2013–

84

pp. 143, *180–3*
Semplice 2012
Table lamp
Oluce, Italy
Blown glass, powder-coated steel spinning
Production 2013–
Elle Decoration Design of the Year, England, 2013; Design Triennial selection, Cooper Hewitt National Design Award, USA, 2015
Permanent Collection: Cooper Hewitt, Smithsonian Design Museum, USA

88

89

90

91

fig. 82
Nut Cracker 2012
Ryohin Keikaku (Muji), Japan
Beechwood
Production 2012–14

fig. 88
Plinths 1 & 2 2012
Benches
Established & Sons, England
Corian
Commission for *Bench Years*, Victoria and Albert Museum, England, 2012

Plinths 1 & 2 are made from a non-porous Corian composite. The design references the museum's courtyard stonework, acting as a juxtaposition between old and new materials.

fig. 85
V Tent 2012
Louis Vuitton
Waxed canvas
Commission for Wallpaper
Handmade, Italy, 2012

The V Tent is a collaboration with Louis Vuitton, who once provided performance tents for early explorers. Our design reinterprets the ancient Viking tent with a simple fabric construction that provides shelter as part of its entrance.

fig. 83
Flower Pot Stand 2012
Terracotta plant pots
Commission from *Süddeutsche Zeitung Magazin* (*SZ*), Germany

fig. 84
Collage 2012
Wall clock
Industrial Facility, England
Reverse clock movement, acrylic, self-adhesive vinyl sheets
Prototype

A reverse clock movement that is normally used for barbershop clocks (so that you can see the time in a mirror) is in this case mounted backwards onto a clear acrylic surface. This results in a distance between the clock movement and the wall, allowing sheets of adhesive film that act as timelines to be stuck behind it. This enables a customer to create a painterly wall collage, blurring the lines of where the clock starts and stops.

pp. 153, *228–9*
Formwork 2012
Desktop storage
Herman Miller, USA
ABS plastic injection moulding, compression-moulded silicone
Production 2014–
iF Gold Award, Germany, 2014; Design Triennial selection, Cooper Hewitt National Design Award, USA, 2015
Permanent Collection: Cooper Hewitt, Smithsonian Design Museum, USA

pp. *80–3*, 133
Radice 2013
Stool
Mattiazzi, Italy
CNC ash, oak woods with natural stains
Production 2013–

pp. 140, *164–9*
Geneva HQ 2013
Security products
Geneva Lab, Switzerland
ABS and styrene plastic
Production 2014–

pp. 141, *174–6*, fig. 90
Tubo 2013
Four-leg and gas-lift chairs
TOG, Italy and Brazil
Bent wire with moulded PET shell
Production 2014–16

pp. 141, *162–3*, fig. 90
Tubo 2013
Table, mobile shelving
TOG, Italy and Brazil
Powder-coated steel rod, marble, urethane and cherry tops
Production 2014–16

A flat-pack table that references the structure of the Tubo chair. The shelving unit comes with wheels.

fig. 91
Branca 2013
Stool
Mattiazzi, Italy
CNC ash, oak woods with natural stains
Production 2014–

A few years after developing the Branca chair [pp. *62–5*], an accompanying stool was designed that used similar CNC production techniques. It has a small reassuring backrest that's also enough to work as a handle or to hang a jacket on.

pp. 146, *198–9*
Busby 2013
Lamp with USB charging
Wästberg, Sweden
Aluminium forging and extrusion, polycarbonate moulding
Production 2015–
Wallpaper Design Award, England, 2016

pp. *42–5*, 120
Together 2014
Lightweight furniture
Herman Miller, USA
Collaboration with Bruce Mau
Paper, steel, HPL
Prototype

pp. *78*, 132
Bottle 2014
Watch
Nava Design, Italy
Anodized aluminium, moulded acrylic, urethane band
Production 2015–18

pp. *100–1*, 137
Remote 2014
Bang & Olufsen, Denmark
Prototype

pp. 149, *212–15*
Run 2014
Table, bench and shelf
Emeco, USA
Cast and extruded aluminium, moulded ABS plastic, cedar, ash and walnut woods
Production 2016–
ICFF Best Collection, USA, 2016; Wallpaper Design Award, England, 2017

Capsule Coffee Maker 2014
Nespresso, Switzerland
Injection-moulded plastic, aluminium casting, steel plate
Prototype

The design of a small, compact coffee maker that worked with a new type of coffee capsule.

fig. 92
Ode 2014
Table and floor lamp
Herman Miller, USA
Spun aluminium, steel tube
Production 2016–
HiP Award, USA, 2016

The Ode lamp was first discussed during the development of the Locale furniture system [pp. *218–21*], becoming its own product line three years later. The idea was to create a freestanding lamp that could also be integrated into a variety of furniture, such as countertops, credenzas, tables or communal desks. There is no physical switch; instead, a light touch to the lamp's pole turns it on or off.

fig. 93, 96
Plex 2014
Modular soft seating
Herman Miller, USA
Foam and fabric, ash, polypropylene plastic
Production 2016–
HiP Award, USA, 2016; Metropolis Likes Award, USA, 2016

Plex is a modular, soft-seating programme for the office that allows for a variety of flexible groupings, using only a small amount of modules. A special hook connects each seat module together. Power can also be integrated, as can wheels or swivel bases.

fig. 100
Figure 2014
Stool
TOG, Italy and Brazil
Beechwood, ABS plastic, height-adjustable mechanism
Prototype

The Figure stool integrates the convenience of gas-lift height adjustment with an archetype of wooden stool design – the utilitarian 'portraiture stool' used by artists. By replacing the common wooden screw adjustment with a gas spring, the stool becomes quicker to adjust for use with height-adjustable tables.

Payment Terminal 2014
Square, USA
ABS plastic, silicone
Prototype

A small wireless device that allows retailers to accept contactless payments.

92

95

93

94

96

fig. 97
ABS Plastic Storage 2014
Desktop accessories
Ryohin Keikaku (Muji), Japan
ABS plastic
Production 2015–

A desktop storage system based on the DIN (Deutsches Institut für Normung) measurement standard, first developed in Germany. DIN enables products to be created that can relate to what has gone before and what may come after. In our example, we designed desk accessories that accommodate the 'A' metric system, already prevalent in Muji. Each piece in the system can stack with every other piece and house a variety of stationery items. Some of the accessories have legs that allow for the creation of raised surfaces, which in turn form further layers of storage.

pp. *48–51*, 122
Beauty as Unfinished Business 2015
Exhibition design and curation
Biennale Internationale Design Saint-Étienne

pp. *84–7*, 134
Dalston 2015
Pendant lamp
Wästberg, Sweden
Anodized die-casting, glass and metal pressings
Production 2016–

pp. 145, *193–5*
OH! Toothbrush 2015
Electric toothbrush
Braun/Oral-B, Germany
Commission for Wallpaper
Handmade, Italy, 2016

pp. *108–11*, 139
Tronco 2015
Chair and table
Mattiazzi, Italy
CNC ash wood with natural stains
Production 2016–

fig. 95
Office Chair 2015
Herman Miller, USA
Injection-moulded glass-filled (30%) nylon, vibration-welded webbing, foam and fabric
Production 2018–

The design is less about innovation and more about introducing simple utility to a chair type that can often be overtly expressive – more machine than furniture. The development process was similar to the low-cost cutlery project [p. 266] some ten years earlier; we deliberately appropriated lines and forms found in bentwood chairs to transform the potentially mechanical appearance into something straightforward and graceful. When seen in numbers, the office atmosphere remains relatively calm.

fig. 98–9
Busby 2015
Exhibition design, twentytwentyone, London
Wästberg, Sweden

Wästberg commissioned us to design a travelling exhibition for the launch of the Busby lamp. Retailer twentytwentyone marked the start with an exhibition held during the London Design Festival in 2015. Large crates made from pinewood provided life-size vignettes, with a bright colour palette that set the tone for imagined settings, such as a workshop, a bedroom or an office. The crates were then packed and shipped to locations in Copenhagen, Stockholm and Oslo.

fig. 94
Plex Tables 2015
Height-adjustable side tables
Herman Miller, USA
Steel, plywood laminate
Production 2017–
HiP Award, USA, 2016; Metropolis Likes Award, USA, 2016

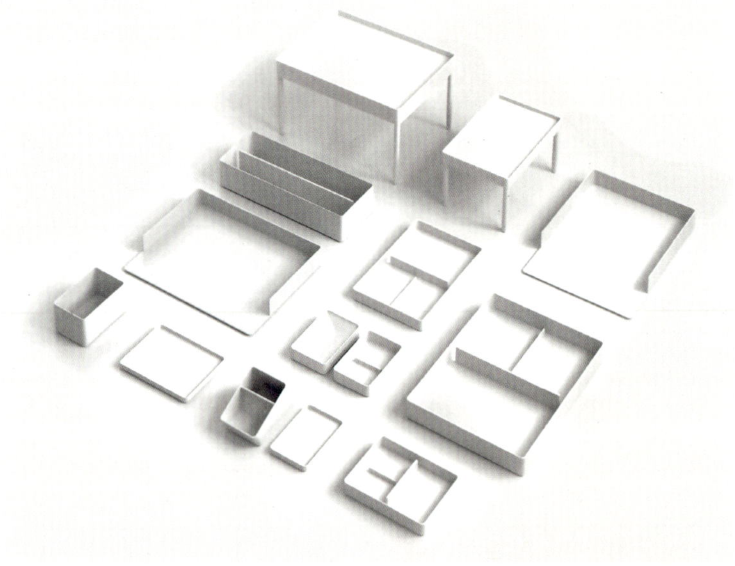

97

98

99

100

101

102

103

104

fig. 102
Platform 2015
Stacking tray
Muuto, Denmark
ABS plastic, white oak veneer
Production 2017–

Like so much of our work, we find ourselves blurring functions between one object and another. In this case, it is a tray, yet its large stubby feet make it feel a little like a small table surface too. These legs also serve another purpose; they allow the trays to stack with one another without the need for an offset lip on its bottom edge.

fig. 111
Run 2016
Exhibition design, Salone del Mobile, Italy, 2016
Emeco, USA

fig. 104
Jar in a Jar 2016
Found glass jars
Commission from *Süddeutsche Zeitung Magazin* (*SZ*), Germany

fig. 109–10
Amazin Apartments 2016
Design Museum, England
Washing machine and refrigerator mechanisms, MDF, steel shelving
Commission for *New Old*, Helen Hamlyn Centre for Design, England, 2017, travelling

The Amazin Apartment concept is the first project designed by Future Facility, Industrial Facility's department dedicated to the Internet of Things. The provocation creates an experience of heightened domestic simplicity, where appliances are reduced to their most essential interface. One example is the washer–dryer, which is positioned at standing height with a shelf below to avoid the need for bending down, with washing powder that is replenished from the rear side. This level of simplicity is made possible because of a service corridor positioned between apartments, which allows service staff to both replenish sundries and also repair or replace appliances should they break, all without the need to enter the apartment.

fig. 107
Easy Bottle 2016
Egg-free mayonnaise, dressing and ketchup bottle
Just Foods, USA
Polyethylene
Production 2018–

fig. 106
Soft Tronco 2016
Upholstered chair
Mattiazzi, Italy
CNC ash wood with natural stains
Production 2017–

pp. 149, *210–11*
Store 2016
Everyday and special editions
Sekisaka Shikki, Japan
ABS plastic, castor aralia wood, urethane and Urushi lacquer
Production 2017–
Commission for Japan Creative

fig. 105
Pastille 2016
Floor, wall and table lamp
Wästberg, Sweden
Powder-coated aluminium casting, injection-moulded polycarbonate
Production 2018–

fig. 108
Street Lighting 2017
Santa & Cole, Spain
Aluminium injection moulding
Production 2018–

A series of street lamp designs that introduce a more domestic and softer appearance.

Pendant Lighting 2017
Santa & Cole, Spain
Moulded glass
Production 2018–

Inspired by dimly lit Spanish tapas bars, a round lamp is hidden underneath a canopy of heavily tinted glass.

Alsik 2017
Pendant lighting
Bitten & Mads Clausen's
Foundation, Denmark
Aluminium extrusion, glass
Prototype

Lighting designed for a new hotel as part of a regeneration programme in the city of Sønderborg.

fig. 103
Reusable Coffee Cup 2017
Cambond, England
Bio-resin, coffee grinds,
polypropylene
Prototype

fig. 101
Facility 2018
Plastic and wood chair
Good Thing, USA
Injection-moulded polypropylene,
OSB board, plywood
Production 2018–

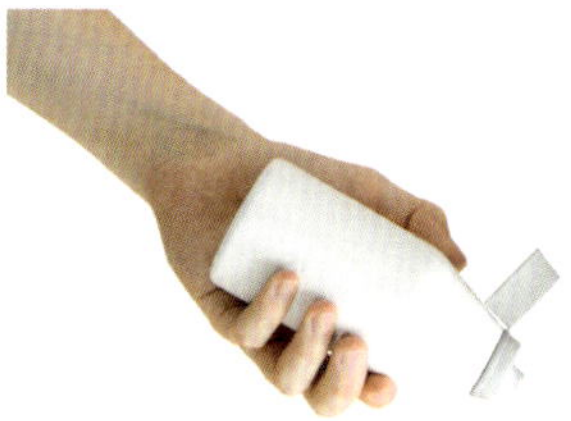

107

108

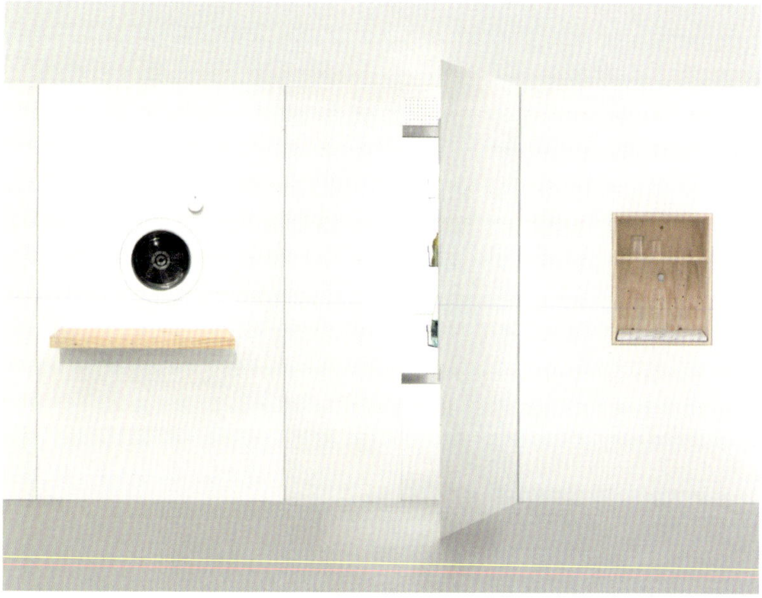

109

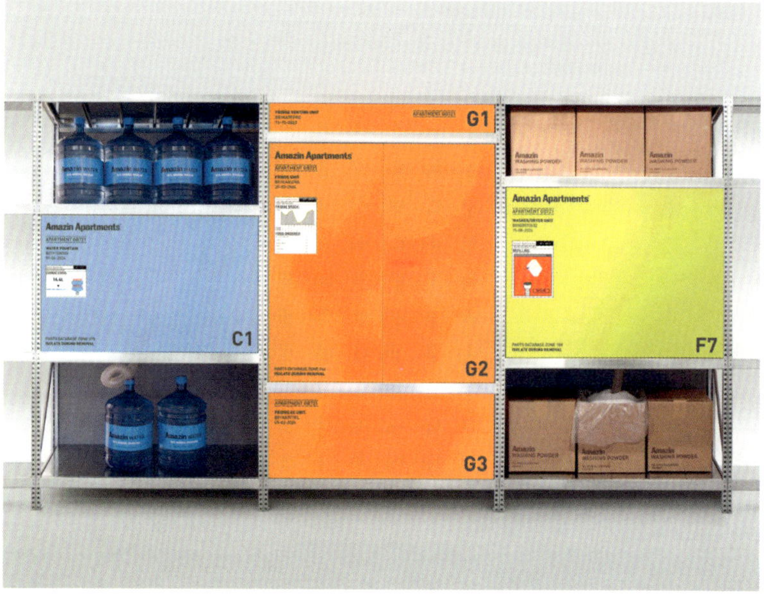

110

105

106

111

Industrial Facility Selected Biography

1961 Colin is born in San Jose, California, to an engineer father and psychologist mother. She moves to Dallas, Texas, before settling in Los Angeles.

1969 Hecht is born in London to a father who owns a record shop in Hackney and a mother who is a nurse.

1972 Colin participates in architect Doreen Gehry Nelson's City Building programme, in which she builds her first model and learns the word architect.

1985–94 Colin completes her Bachelor of Arts in art history under John Tagg and Charles Jencks at UCLA; she also works as a research assistant to Margaret Crawford in the school's Richard Neutra archive. Colin receives her Masters in Architecture from SCI-Arc, Los Angeles, under Michael Rotondi.

1987–92 Hecht enrols at the Central School of Design, followed by the Royal College of Art (RCA), both in London, under Daniel Weil, where he receives his Masters in Industrial Design.

1993 After working briefly with architect David Chipperfield, Hecht moves to Tel Aviv to work with architects he met at the Architectural Association (AA), London. He travels to Boston to work with Harvard University professor Frank Stone, sponsored by Bill Moggridge.

1994 Hecht joins IDEO's San Francisco office, working with, amongst others, Tim Brown, Naoto Fukasawa, Sigi Moeslinger and Matt Rohrbach.

1995 Colin works with Los Angeles-based artist Mike Kelley and builds his *Educational Complex* artwork, now in the collection of the Whitney Museum of American Art, New York.

1996 Fukasawa establishes an IDEO office in Tokyo and invites Hecht to join him; Takeshi Ishiguro joins them a year later.

1998 Colin interviews Pierre Koenig and Julius Shulman for Koenig's first monograph; she then moves to London to commission books on architecture and design for Phaidon Press.

1999–2002 Hecht returns to London to work as head of design at IDEO; he meets Colin and they start to collaborate. Colin is invited to teach students of the Architecture and Design Studios with Clive Sall at the RCA and the AA.

2002–3 Colin and Hecht establish Industrial Facility in London; their first office is on Golden Lane in Clerkenwell, shared with Anthony Dunne and Fiona Raby. They publish their first book, *Things That Go Unseen*. Projects include furniture, electronics, consumer products, fashion and interiors. They become retained designers to Muji and Hecht presents lectures at Toto Gallery MA, Tokyo; Kiel University, Germany; Design Academy Eindhoven, the Netherlands; and the RCA, London.

2004 Industrial Facility publish their second book, *Product as Landscape*, and start to work with Taylor's Eye Witness at the request of the British Design Council. Kenya Hara invites them to contribute to his *Haptic* exhibition in Tokyo. Lectures are given at the Centre Pompidou, Paris and Stuttgart University, Germany.

2005–6 The Israel Museum commissions the exhibition, *Found/Made/Thought, The Work of Industrial Facility*. The British Council includes selected studio work in the exhibition *Import Export*, first shown in Helsinki.

2007 Herman Miller invite Colin and Hecht to become design advisors in the USA, where they meet Bruce Mau and Susan Lyons. Hecht forms Platform 12 in Design Products at the RCA. The British Council includes selected work in the exhibition *Love & Money* and Hecht gives a corresponding lecture in Taipei. Colin and Hecht receive Material ConneXion's honorary award in New York.

2008 Hecht is conferred as a Royal Designer for Industry by the Royal Society of Arts (RSA), London, and is short-listed for the Prince Phillip Design Prize.

2009 Deyan Sudjic commissions the studio for two adjacent exhibitions at the Design Museum, London, *Industrial Facility: Some Recent Projects* and *Under a Fiver*. Their online shop www.retailfacility.co.uk is launched. *ICON* magazine editors Justin McGuirk and Johanna Agerman commission a monthly column called 'Under a Fiver: Recent Acquisitions'. Hecht gives a lecture at Design East, Osaka.

2010 Industrial Facility's book *Usefulness in Small Things* is published. Branca, a robotically produced wooden chair for Mattiazzi, debuts at Salone del Mobile, Milan and recent RCA graduate, Juriaan Booij, is commissioned by Hecht and Colin to make a film about it. Lectures are given at Cranbrook Academy of Art, Bloomfield Hills, Michigan; HfG Karlsruhe, Germany; and the RSA, London.

2011 Hecht becomes visiting professor at HfG Karlsruhe and his book of student work, *Placidity in Technology*, is published. Group exhibitions are held in London, Beijing and Paris. Hecht and Colin's Branca chair receives the Design Museum's Design of the Year award for furniture. The office moves to Britton Street in Clerkenwell, London.

2012 Lectures are given at Mudam, Luxembourg. Group exhibitions are held in Chicago and London.

2013 Colin and Hecht visit Finland to lecture at the Alvar Aalto design seminar. Projects begin with Wästberg in Sweden.

2014 Biennale Internationale Design Saint-Étienne commissions the exhibition *Beauty as Unfinished Business*. The book *Locale* is published by Herman Miller.

2015 Colin is conferred as a Royal Designer for Industry by the RSA, the first female product designer to be awarded since its formation. Lectures are given in Portugal for the Monocle Quality of Life conference, and in San Francisco. Small exhibitions for the Busby lamp are held in Sweden and at twentytwentyone in London.

2016 Colin and Hecht launch Future Facility, a branch of their studio that deals with the future of products and the Internet of Things. A workshop is held with the Kyoto Design Lab in Japan, coinciding with research for Japan Creative. Coexistence in London hosts the launch of the Run collection for Emeco.

2017 Colin is invited to teach and lecture at ECAL, Lausanne in the Masters of Product Design programme. She gives lectures in San Francisco and for the IKEA Foundation in Switzerland.

1

2

3

4

5

1 Inside Industrial Facility in Clerkenwell, London, 2012

2 Gregg Buchbinder shows Sam around the Emeco factory in Hanover, Pennsylvania, 2014

3 Sam photographed by Maria Ziegelböck for *10* magazine, 2012

4 Setting up Locale models for photographer Gerhardt Kellermann, 2013

5 'A handstand a day' as architect Albert Frey used to preach. Kim and Sam practice with their eldest son, 2004

6

9

12

7

8

10

11

6 Sam and Philipp von Lintel drop into a crisp factory next to Emeco in Hanover, Pennsylvania, 2014

7 Observing the Urushi lacquerware process in Sabae, Japan, arranged by Wataru Kumano and Japan Creative, 2016

8 Chantal Hamaide of *Intramuros*, with architect Philippe Boisselier, Sam and Kim after building their exhibition at Maison et Objet, Paris, 2003

9 Sam with fellow Royal College of Art tutors Durrell Bishop and André Klauser, walking the streets of London with artist Richard Wentworth, 2012

10 The first office of Industrial Facility in Golden Lane, London, 2003

11 The late Alex Ward, curator at the Israel museum, catches Sam and Ippei Matsumoto as they build Industrial Facility's exhibition, 2005

12 Kim relaxes in a Branca chair during Mattiazzi's photoshoot in Regent's Park, London, 2017

13

17

14

18

15

19

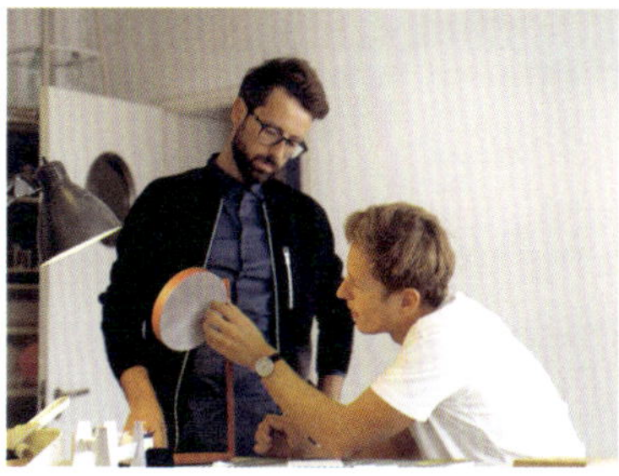

16

13 Herman Miller engineers with Sam in Michigan during development of Locale, 2011

14 Kim building Mike Kelley's artwork *Educational Complex*, 1995

15 Christoph Mack captures Sam and Kim for Mattiazzi's 'Together Apart' campaign, designed by Lambl Homburger, 2017

16 Assistants Luca Corvatta and Romain Voulet discuss the development of a lamp for Wästberg, 2017

17 Outside the Industrial Facility studio in Clerkenwell, London, 2012

18 Royal College of Art students visit the studio, 2011

19 Sasaki Hidetoyo photographs Sam and Naoto Fukasawa in his studio in Tokyo, 1997

20

22

23

21

24

20 Artist Martin Creed and Sam brought together in conversation by Johanna Agerman during her time at *ICON* magazine, 2008

21 Sam and Kim demonstrating the Together table's versatility at Herman Miller, 2014

22 Photographer Gerhardt Kellermann captures Kim at the Clerkenwell office, 2013

23 Kim with Don Goeman, executive vice president of research and design at Herman Miller, 2011

24 Kim with her architecture students from the Royal College of Art, setting up their exhibition at the Witte De With in the Netherlands, 2000

Industrial Facility Selected Exhibitions

1999 *Identity Crisis*, The Lighthouse, Glasgow, Scotland

2000 *Under a Fiver*, Design Museum, London, England; *Play-Use*, Witte de With, Rotterdam, the Netherlands

2001 *Great Expectations*, Department for International Trade, Shanghai, China, travelling; *Workspheres*, Museum of Modern Art, New York, USA

2002 *Spot-on*, Gallery Haute Definition, Paris, France; *Skin*, Cooper Hewitt, Smithsonian Design Museum, New York, USA

2003 Muji Exhibition, during Salone del Mobile, Milan, Italy; *Artist and Engineer*, Maison et Objet, Paris, France; *The Future and Muji*, Toto Gallery MA, Tokyo, Japan; *European Design Show*, Design Museum, London, England

2004 *Import Export*, British Council, travelling; *Haptic*, Spiral Gallery, Tokyo, Japan, travelling; Architectural Biennale, Beijing, China

2005 *Found, Made, Thought, The Work of Industrial Facility*, Israel Museum, Jerusalem; *European Design Show*, Design Museum London, England; *Under a Tenner*, Design Museum, London, England; *Objet du Conversation*, Le Bon Marché and Galerie kreo, Paris, France; *Value for Money*, Droog BV, during Salone del Mobile, Milan, Italy

2006 *Love & Money*, British Council, Taipei, Taiwan, travelling; Biennale Internationale Design Saint-Étienne, France; *Football Fever*, Design Museum, London, England; *Hearwear*, Victoria and Albert Museum, London, England

2007 *Some Recent Projects*, Design Museum, London, England

2008 *Design Cities*, Design Museum, London, England; *Design and the Elastic*, Museum of Modern Art, New York, USA

2009 *Super Contemporary*, Design Museum, London, England; *Super Normal*, twentytwentyone, London, England, travelling

2010 *New Simplicity*, Brompton Road, London, England; Design East, Megamag, Osaka, Japan

2011 *Turn, Twist and Branch Off*, Aram Store, London, England; *Power of Making*, Victoria and Albert Museum, London, England; Beijing International Design Triennial, China; Paris Photo, Grand Palais, France; Biennale Interieur, Kortrijk, Belgium; *Designs of the Year*, Design Museum, London, England

2012 *Less and More, The Work of Dieter Rams*, San Francisco Museum of Modern Art, USA; *Extraordinary Stories about Ordinary Things*, Design Museum, London, England; *Table Items*, Art Institute of Chicago, USA; *Bauhaus Influences*, Aram Gallery, London, England; *Wallpaper Handmade*, during Salone del Mobile, Milan, Italy; *Bench Years*, Victoria and Albert Museum, London, England; *Muji Product Fitness 80%*, Design Museum, London, England; Beijing Design Week, China

2014 *GREAT*, Sakip Sabanci Museum, Istanbul, Turkey; *Useful and Beautiful*, Geffrye Museum, London, England; *Guild*, British Council, Cape Town, South Africa

2015 *Family Matters*, Die Neue Sammlung, Munich, Germany; *Serial Beauty*, Biennale Internationale Design Saint-Étienne, France; *Houselife*, Musée des Arts décoratifs et du Design (MADD), Bordeaux, France; *Beauty as Unfinished Business*, Biennale Internationale Design Saint-Étienne, France; *Beauty*, Cooper Hewitt, Smithsonian Design Museum, New York, USA, travelling; *Recent Works*, Skridskopaviljongen, Stockhom, Sweden; *Busby*, twentytwentyone, London, England

2016 *Safe and Sound*, Musée de design et d'arts appliqués contemporains, Lausanne, Switzerland; *Wallpaper Handmade*, during Salone del Mobile, Milan, Italy

2017 Singapore Design Week, Singapore; Japan Creative, during Salone del Mobile, Milan, Italy; Łódź Design Festival, Poland; *New Old*, Design Museum, London, England, travelling

Contributor biographies

Durrell Bishop is a British designer who experiments with products, installations and hardware interfaces. His research and development bring unusual products to market, including the recent Line-us drawing robot. He obtained a Masters in both industrial design and interaction design from the Royal College of Art and has taught regularly there. He worked at IDEO, Interval Research Corporation and Berg; and co-founded studios Itch and Luckybite.

Alain de Botton is a British author who was born in Zurich, Switzerland. His books discuss various contemporary subjects and themes that are relevant to everyday life. His works include *The Architecture of Happiness* (2006) and *Essays in Love* (1993), which has sold more than two million copies. In 2008 he co-founded The School of Life, followed by Living Architecture, in 2009. In 2015 he was awarded The Fellowship of Schopenhauer by the Melbourne Writers Festival.

Naoto Fukasawa is a Japanese industrial designer who graduated from Tama Art University in 1980, where he is now a professor. After acting as the head of IDEO's Tokyo branch, he established his own office in 2003. His works include Muji's CD player, the ±0 brand of household appliances and designs for companies including B&B Italia, Magis, Maruni and Artemide. His writings include *The Outline: The Unseen Outline of Things* (2009), and he is co-director of 21_21 DESIGN SIGHT, Japan's first design museum.

Don Goeman is Herman Miller's executive vice president of R&D and is responsible for the research, design and development of new product programmes. He joined the company in 1980, and during the past thirty-seven years has held a variety of new product design and development positions. Most noteworthy are collaborations with industrial designers that have led to production projects, including the Aeron chair and the Locale Office System. Don is a graduate of the University of Michigan, with a BSE in Mechanical Engineering.

Sam Grawe was appointed global brand director of Herman Miller in 2012, and is responsible for its corporate archives, communications and media, while also overseeing the creation of content for print, digital and exhibitions. From 2006–11, he served as the editor-in-chief of *Dwell* magazine, being part of the team that garnered the National Magazine Award and the position of finalist in the National Design Awards.

Peter Kapos trained in both fine art and philosophy. He is a curator and writer, a lecturer in art and design at Camberwell College of Arts, University of the Arts London, and the creative director of Systems Studio graphic design office. His projects include *The Ulm Model*, an exhibition of HfG Ulm design held at Raven Row exhibition centre in London, and writing the Foreword for the republication of *Braun+Design Collection* (1999).

Anniina Koivu graduated as an architect from Aalto University, Finland, before becoming design editor of *Abitare* magazine from 2007–11. From 2011–14 she was director of research at Vitra, and is now head of theory for the Master's programmes in product design, photography and typography at ECAL. She has contributed to a number of books on design and is editor-in-chief of *Interwoven*, an online magazine she launched with Kvadrat in 2016.

Bruce Mau is a Canadian designer whose creative talent extends beyond graphics and into the worlds of architecture, art, conceptual philosophy and socio-environmental design. He designed and co-authored the influential book *S,M,L,XL* with Rem Koolhaas in 1995. From 1985–2010, he was creative director of Bruce Mau Design (BMD). In 2010 he co-founded the Massive Change Network with Bisi Williams and in 2015 events company Freeman appointed him chief design officer. He is Distinguished Fellow at the Segal Design Institute, McCormick School of Engineering at Northwestern University.

Eizo Okada was born in Japan and is currently professor of the Kyoto Institute of Technology as well as being design director of S&O Design. He completed his PhD studies at Chiba University, and has directed various projects in design and architecture, while carrying out research that focuses on social design processes. He edits the design news website, dezain.net, and has written extensively for design and media magazines.

Leon Ransmeier is an American industrial designer. A graduate of the Rhode Island School of Design, he founded his New York-based studio, Ransmeier Inc. in 2010. He has since designed furniture and products for clients including Arita, HAY, Maharam, Mattiazzi and Herman Miller, for whom he has co-edited a book about the company's history.

Duncan Riches studied modern history and politics at London Guildhall University, before going on to work in the design industry as a writer, consultant and strategist. He set up his own content consultancy in 2007 and has worked with leading companies and designers, including SCP, HAY, Wrong London, Established & Sons, Johnson Naylor and tokyobike.

Oli Stratford was born in Yorkshire, studied philosophy in Cambridge and journalism in London. He has been the editor-in-chief of *Disegno* since September 2016, having previously served as the journal's online and deputy editor. Outside *Disegno*, examples of his journalism, interviews, essays and fiction may be read in books published by König Books and Gestalten, among others.

Deyan Sudjic was born in London and trained to be an architect in Edinburgh. He has worked as a critic, editor and curator in Milan, Glasgow, Vienna and Istanbul, and has been director of the Design Museum since 2006. He has written a number of books, including a biography of Ettore Sottsass and a catalogue raisonné on Shiro Kuramata. His most recent book, *The Language of Cities*, was published in 2017.

Magnus Wästberg was born in Helsingborg, southern Sweden, to a father who worked in the lighting industry. He studied business administration at Lund University, before starting his own lighting company in 2008. His ambition was, and still is, to restore the delicate balance between Man and Light. Over the last decade, Wästberg has cooperated with some of the world's most distinguished architects and designers, receiving more than fifty international design awards.

Index

Page numbers in *italics* refer to illustrations